Guide to the Marine Stations of the North Atlantic and European Waters

Part 2

Mediterranean Basin

compiled on behalf of
The Royal Society's Naples Zoological Station Committee
by J. E. Webb

ISBN 0 85403 079 4

The Royal Society
6 Carlton House Terrace
London, SW1Y 5AG

C O N T E N T S

CONTENTS (continued)

PREFACE

The intention of this guide is to help scientists select marine stations with facilities for their research.

Prospective visitors want to know about the types of shore in the neighbourhood, the animals and plants of special interest that can be collected easily and also the range of boats, laboratories and equipment that would be made available for their use.   In addition, as many visitors to marine laboratories combine research with a vacation, information on living accommodation at the Station or nearby for themselves and their families is helpful.

The information has been given in a reply to a questionnaire by the staff at each station.   This method ensures some uniformity of presentation while photographic reproduction of the edited questionnaires means that revision when needed will not be difficult.   The guide was not meant to be comprehensive, but was intended to include only those stations that want to have visitors.   We have written to all stations known to us.   Not all have replied and some may have been overlooked, but it is hoped that such omissions can be rectified in the first revision.

It is hoped that the first part of the guide on the marine stations of northern Europe and the east Atlantic coast and this second part on the Mediterranean region will be followed by another on the west Atlantic coast and the Caribbean.

August 1975

J. E. Webb
Westfield College
University of London

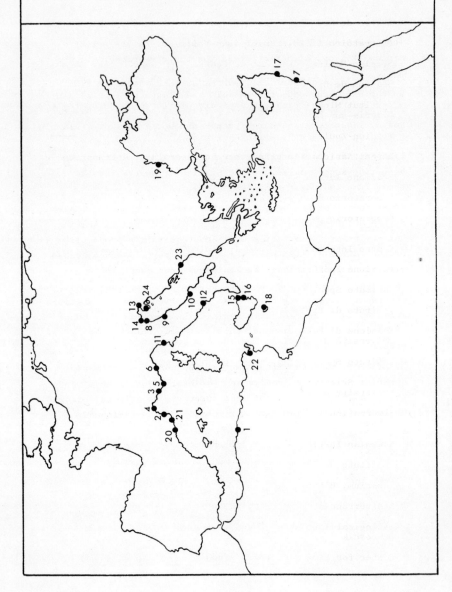

K E Y   T O   M A P

1   Centre de Recherche Océanographique et de Pêche, Alger

2   Laboratoire Arago, Banyuls-sur-Mer

3   Station Marine d'Endoume et Centre d'Océanographie, Marseille

4   Station de Biologie Marine et Lagunaire, Sete

5   Institut Michel Pacha, Laboratoire maritime de Physiologie, Tamaris-sur-Mer

6   Station Zoologique, Villefranche-sur-Mer

7   Israel National Institute for Oceanography and Limnology, Haifa

8   Stazione Idrobiologica, Chioggia

9   Laboratorio di Biologia marina e Pesca, Fano

10   Laboratorio per lo Sfruttamento Biologico delle Lagune CNR, Lesina

11   Centro Interuniversitario di Biologia Marina, Livorno

12   Stazione Zoologica di Napoli, Napoli

13   Istituto Sperimentale Talassografico 'F.Vercelli', Trieste

14   Istituto di Biologia del Mare, Venezia

15   Stazione di Biologia Marina dell'Istituto di Zoologia dell' Università di Messina, Messina

16   Istituto Sperimentale Talassografico, Messina

17   Marine Science Program, Department of Biology, American University of Beirut, Beirut

18   International Ocean Institute, Royal University of Malta, Msida

19   Rumanian Institute for Marine Research, Constantza

20   Instituto de Investigaciones Pesqueras, Barcelona

21   Aquarium, Blanes

22   Mediterranean Marine Sorting Center, Khereddine

23   Biological Institute, Yugoslav Academy of Sciences and Arts, Dubrovnik

24   Center for Marine Research 'Rudjer Bošković' Institute, Rovinj

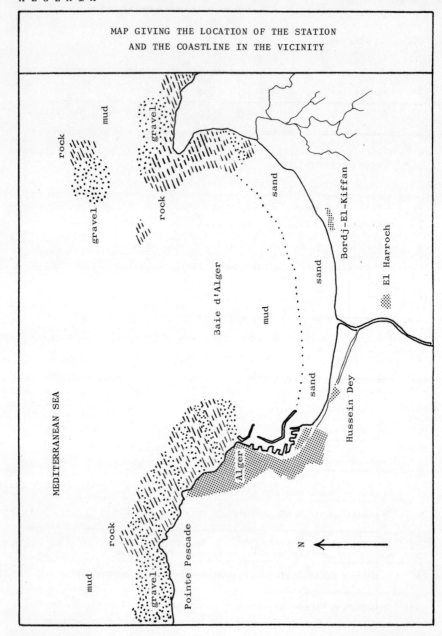

MAP GIVING THE LOCATION OF THE STATION
AND THE COASTLINE IN THE VICINITY

A L G E R I A

Name and address of the Station	Centre de Recherche Océanographique et de Pêche BP 90 Jetée Nord Alger Algerie
	Telephone 62-73-11 62-66-59
Director	Mr. R. Semroud
Affiliation	Organisme National de la Recherche Scientifique Ministère de l'Enseignement Superieur et de la Recherche Scientifique

NUMBER OF PERMANENT STAFF	
Scientific	2 Phytoplanktologists 2 Zooplanktologists 4 Ichthyologists 2 Zoologists 2 Fishery staff
Technical	5
Other	3

ROUTES OF ACCESS	
Air	Airport 'Dar El Beida' Alger
Rail	—
Road	From Spain-Morocco-Algeria through Gibraltar Detroit
Sea	Port d'Alger from Spain and France (mostly Marseilles)

9

MAJOR ACCESSIBLE ECOSYSTEMS	
Intertidal	Yes
Shallow sea	Yes
Ocean	Yes
Tidal range	Negligible

COLLECTING AND HYDROGRAPHIC FACILITIES	
on shore	Yes
from boats	Yes, dredges, plankton nets, Hansen bottles, etc.
Diving (including supply and/or refill of cylinders)	Yes, but no equipment loaned
Collecting service for live material provided by the Station	Yes, but not always possible
Hydrographic data	Yes

SPECIAL ORGANISMS AND LOCALITIES THAT ARE A FEATURE OF THE STATION		
name	abundance	season
None given		

LABORATORIES OPEN TO VISITORS	
General laboratories	Yes
Physiology laboratories	—
Biochemical laboratories	—
Research rooms for visitors	Yes
Wet sorting room	—
Experimental aquarium	Can be constructed if necessary
Controlled environment rooms	Yes
Photographic dark room	Yes
Workshop for general repairs to apparatus	Yes

LABORATORY SERVICES	
Electricity - voltage, DC/AC, frequency, phase	110 V, 220 V
Gas	Yes, natural gas
Running seawater	Yes
Compressed air (pumps)	Yes
Computer link	At the Nuclear Center
General glassware	Yes
Chemicals	Yes
Balances (analytical)	Yes
Balances (torsion)	Yes
Refrigerators	Yes
Ovens	Yes
Microscopes (compound)	Yes
Microscopes (dissecting)	Yes
Electron microscope	At the Algiers University
Centrifuges	Yes
Special apparatus	Histological equipment
Library	Yes

A L G E R I A

LIVING ACCOMMODATION FOR VISITORS AND THEIR FAMILIES	
At the Station	Yes, three single bedrooms
Restaurant	Yes, near the Station
Local Hotels	Yes
Camp Sites	——
Car parking	Yes

APPROXIMATE CHARGES	
Laboratory	No charge
Boat	
Specimens	
Living accommodation at the Station	
Meals at the Station	Meals are not served
Hotels (range/person/night)	——
Other	——

PUBLICATIONS GIVING INFORMATION OF VALUE TO VISITORS

The Station's review 'Pelagos' contains some information about hydrography and zooplankton.

ADDITIONAL INFORMATION SUPPLIED BY THE STATION

The Station welcomes many international scientists. Applicants wishing to visit should write to the Director of the Centre or to the Directeur General de la Recherche Scientifique, 1 Rue Bachir Attar, Alger, and give their proposed scheme of research.

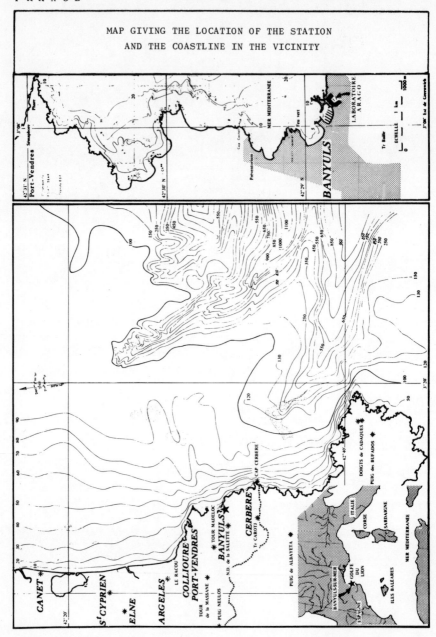

MAP GIVING THE LOCATION OF THE STATION
AND THE COASTLINE IN THE VICINITY

F R A N C E

Name and address of the Station	Laboratoire Arago 66650 Banyuls-sur-Mer France
	Telephone 38.30.09
Director	Professeur Pierre Drach
Affiliation	Université Pierre et Marie Curie 4 place Jussieu 75230 Paris Cedex 05

NUMBER OF PERMANENT STAFF	
Scientific	1 Professeur 5 Maîtres-assistants 3 Assistants
Technical	40 dont 7 affectés à l'aide à la recherche 12 marins
Other	35 Chercheurs permanents

ROUTES OF ACCESS	
Air	Perpignan Llabanere
Rail	Port-Vendres (7km de Banyuls) Banyuls en été
Road	N 114 (35 km de Perpignan)
Sea	——

MAJOR ACCESSIBLE ECOSYSTEMS	
Intertidal	Réduit car méditerranéen
Shallow sea	Tous les types de substrats (durs et meubles) Très beau 'coralligène'
Ocean	Canyons sous-marins (150 - 600 m de profondeur)
Tidal range	35 cm

COLLECTING AND HYDROGRAPHIC FACILITIES	
on shore	2 véhicules permettant l'accès à tous les points du littoral 2 zodiacs 1 barque
from boats	Un navire de 17m 50 (Pr. Lacaze-Duthiers) Une vedette de 12m 50 (Nereis) permettant d'effectuer des récoltes benthiques (chaluts, dragues, bennes, carottiers) et pélagiques (filets Clark-Bumpus) ainsi que les mesures hydrologiques classiques depuis la côte jusqu'à 600 - 800 m environ
Diving (including supply and/or refill of cylinders)	Service de plongée (2 plongeurs professionnels) Met à la disposition des chercheurs 20 bouteilles et détendeurs. Recharge gratuite. Les plongeurs doivent avoir une assurance couvrant les usagers de la plongée professionnelle et apporter leur matériel personnel (combinaison palmes, masque, tuba, ceinturon)
Collecting service for live material provided by the Station	Très grande possibilité de récolte de matériel vivant varié (signaler le matériel nécessaire) Service d'envoi de matériel fixé (matériel d'enseignement et de recherches) - Catalogue sur demande
Hydrographic data	Nombreuses références bibliographiques récentes disponibles Matériel d'hydrographie classique disponible Salinomètre électronique

SPECIAL ORGANISMS AND LOCALITIES THAT ARE A FEATURE OF THE STATION		
name	abundance	season

Biotopes particuliers:
  Coralligène
  Trottoir à Lithophyllum
  Sables à Amphioxus		

Animaux
  Paracentrotus lividus
  Arbacia lixula
  Amphioxus
  Céphalopodes divers
	très abondant
	très abondant
	abondant
	variés et abondants
		mars, avril,
		septembre, octobre

Méroplancton	riche	Printemps

Matériel disponible très varié du fait de la diversité des côtes et de la grande gamme des profondeurs accessibles par les moyens du Laboratoire.

LABORATORIES OPEN TO VISITORS	
General laboratories	Oui
Physiology laboratories	Non
Biochemical laboratories	Non
Research rooms for visitors	Oui
Wet sorting room	Oui
Experimental aquarium	Oui
Controlled environment rooms	Oui
Photographic dark room	Oui - sous la responsabilité d'un photographe professionnel
Workshop for general repairs to apparatus	Oui

LABORATORY SERVICES	
Electricity – voltage, DC/AC, frequency, phase	110 – 125 V AC 50 Hz 210 – 220 V AC 50 Hz mono et triphasé
Gas	Oui
Running seawater	Oui
Compressed air (pumps)	Oui
Computer link	Non
General glassware	Oui
Chemicals	Oui
Balances (analytical)	Oui
Balances (torsion)	Electromagnétiques
Refrigerators	Oui
Ovens	Oui
Microscopes (compound)	35 (Wild, Leitz, Zeiss)
Microscopes (dissecting)	30 (Wild, Leitz)
Electron microscope	Non
Centrifuges	de paillasses ultracentrifugeuse MSE superspeed réfrigéré
Special apparatus	Technicon, Lyophilisateur, Spectrophoto-mètres, Fluorimètre, Ultramicrotomes manuels et automatiques, Microtomes, etc. Salle de culture stérile
Library	Très importante bibliothèque spécialisée en biologie marine (1500 périodiques par abonnement ou échanges; 150.000 tirés à part; nombreux traités)

F R A N C E

LIVING ACCOMMODATION FOR VISITORS AND THEIR FAMILIES	
At the Station	54 lits répartis en chambres à 2 ou 3 lits Nécessaire de réserver pour la saison (mars à octobre). En principe les familles avec enfants ne sont pas logées
Restaurant	Restaurant à prix modérés fonctionnant de Février à fin Novembre
Local Hotels	S'adresser au Syndicat d'Initiative
Camp Sites	Oui
Car parking	Oui

APPROXIMATE CHARGES	
Laboratory	Free
Boat	Free
Specimens	Free - lorsque les chercheurs sont sur place
Living accommodation at the Station	2 à 7 frs suivant le grade universitaire
Meals at the Station	Pension complete: 15 frs
Hotels (range/person/night)	Prix suivant classe
Other - pour les stages	63,20 F par jour pour l'ensemble (30 pers. min.) 5,10 F par étudiant et par jour

FRANCE

PUBLICATIONS GIVING INFORMATION OF VALUE TO VISITORS

Les recherches entreprises au Laboratoire Arago sont
en partie publiées dans le périodique 'Vie et Milieu'.

Le rapport d'activité peut être fourni sur demande.

Les demands d'information sont à adresser à:

> Laboratoire Arago
> Service Administratifs
> 66650 Banyuls-sur-Mer
> France

ADDITIONAL INFORMATION SUPPLIED BY THE STATION

Le Laboratoire Arago est un laboratoire de recherche
et d'enseignement.   Il a une importante fonction
d'accueil, comme la plupart des laboratoires de terrain.
A titre d'exemple, en 1974, ont été effectués:
33 stages dont 18 français et 15 etrangers, suivis pas
562 étudiants.
171 séjours de recherches, dont 90 par des français et
81 par des chercheurs étrangers.

L'ensemble des prestations à caractère scientifique
fournis par le laboratoire est gratuit.   Cependant,
en cas d'utilisation de produits chers (radioactifs,
fixateurs pour microscope électronique), il est
demandé d'apporter ses propres fournitures.

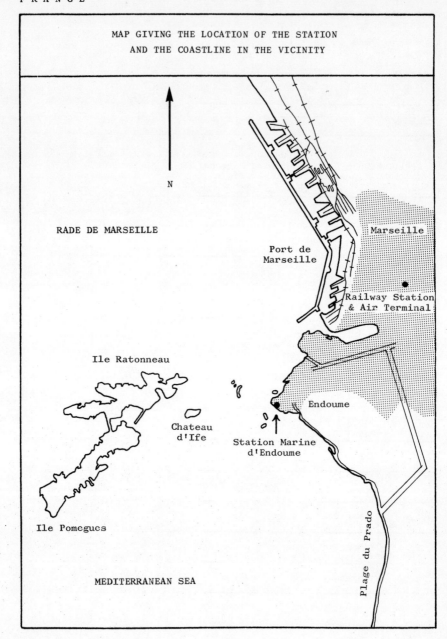

MAP GIVING THE LOCATION OF THE STATION
AND THE COASTLINE IN THE VICINITY

N

RADE DE MARSEILLE

Port de
Marseille

Marseille

Railway Station
& Air Terminal

Ile Ratonneau

Chateau
d'Ife

Endoume

Station Marine
d'Endoume

Ile Pomegues

Plage du Prado

MEDITERRANEAN SEA

Name and address of the Station	Station Marine d'Endoume et Centre d'Océanographie Rue Batterie des Lions 13007 Marseille France
	Telephone 52 12 94

Director	Prof. J. M. Pérès

Affiliation	University of Aix-Marseille (II) Centre National de la Recherche Scientifique

NUMBER OF PERMANENT STAFF	
Scientific	About 90
Technical	About 40
Other	Administrators 5

ROUTES OF ACCESS	
Air	Marignane airport 32 Km
Rail	Marseilles
Road	Highway 6-7 (from Paris)
Sea	Marseilles

F R A N C E

MAJOR ACCESSIBLE ECOSYSTEMS	
Intertidal	Negligible
Shallow sea	Yes
Ocean	Yes
Tidal range	Max. 0.5 m

COLLECTING AND HYDROGRAPHIC FACILITIES	
on shore	Yes
from boats	Yes
Diving (including supply and/or refill of cylinders)	Yes
Collecting service for live material provided by the Station	No
Hydrographic data	No regular readings taken

F R A N C E

SPECIAL ORGANISMS AND LOCALITIES THAT ARE A FEATURE OF THE STATION		
name	abundance	season
None given		

LABORATORIES OPEN TO VISITORS	
General laboratories	Yes
Physiology laboratories	Minimum facilities only
Biochemical laboratories	Yes
Research rooms for visitors	7 rooms
Wet sorting room	No
Experimental aquarium	Yes (some with temperature regulation)
Controlled environment rooms	Yes and cold rooms
Photographic dark room	Yes (2)
Workshop for general repairs to apparatus	Yes

LABORATORY SERVICES	
Electricity - voltage, DC/AC, frequency, phase	110 - 220 V AC 50 Hz
Gas	Main supply in three of the four buildings In the fourth 'camping gas'
Running seawater	In all buildings and on all floors
Compressed air (pumps)	No
Computer link	Yes (at the University 1.5 Km)
General glassware	Minimal
Chemicals	Yes
Balances (analytical)	Yes
Balances (torsion)	No
Refrigerators	Yes
Ovens	Yes
Microscopes (compound)	Yes
Microscopes (dissecting)	Yes
Electron microscope	Yes
Centrifuges	Yes
Special apparatus	A wide range such as: mass spectrometer, amino acid auto-analyzer, spectro-photometers (5)
Library	Yes - in all fields of Marine Science Also library at the University

F R A N C E

LIVING ACCOMMODATION FOR VISITORS AND THEIR FAMILIES	
At the Station	Yes - 2 double rooms, 5 single rooms
Restaurant	Lunch only Monday to Friday except from July 13 to September 1
Local Hotels	Yes
Camp Sites	5 Km
Car parking	Yes

APPROXIMATE CHARGES	
Laboratory	No charge except for glassware, expensive chemicals and boats required for special purposes outside the general needs of the Station (boats 150 - 1800 FF according to size)
Boat	
Specimens	
Living accommodation at the Station	10 FF/day/person
Meals at the Station	5 FF
Hotels (range/person/night)	A rather cheap hotel 150 m from the laboratory. Many hotels downtown
Other	——

PUBLICATIONS GIVING INFORMATION OF VALUE TO VISITORS

The Marine Station has its own journal 'Tethys'

ADDITIONAL INFORMATION SUPPLIED BY THE STATION

All fields of research in Marine Science covered except
physical oceanography per se and geophysics.

Some research groups are located at the University
campus (15 Km from the Station).

Due to the vacations of technicians the activity of
the laboratory is greatly reduced between July 12 and
the beginning of September.

All applications for working in the laboratory should
be addressed to the Director.

Visitors should arrive between Monday 07.00 am and
Saturday 12.00 and not at weekends.

NOTES

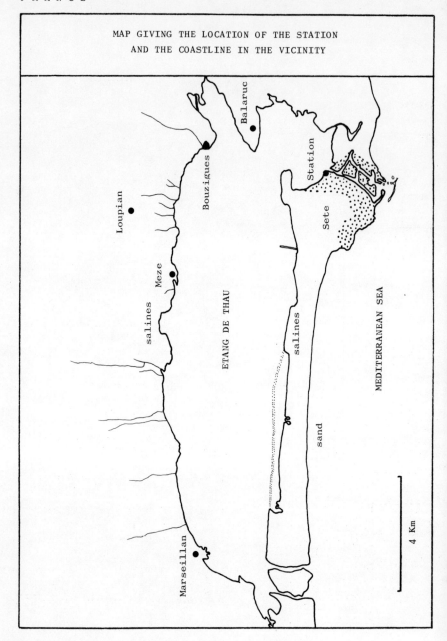

MAP GIVING THE LOCATION OF THE STATION
AND THE COASTLINE IN THE VICINITY

F R A N C E

Name and address of the Station	Station de Biologie Marine et Lagunaire Quai de Bosc Prolongé Sete 34200 France
	Telephone (67) 74 36 70
Director	Professeur Jean Paris
Affiliation	Université des Sciences et Techniques du Languedoc

NUMBER OF PERMANENT STAFF	
Scientific	5
Technical	4
Other	2

ROUTES OF ACCESS	
Air	Aerodrome Frejorgues/Montpellier (28 Km)
Rail	Sete Railway station (SCNF)
Road	Autostrade Paris-Narbonne 8 Km
Sea	Commercial harbour

F R A N C E

MAJOR ACCESSIBLE ECOSYSTEMS	
Intertidal	Marine and brackish water
Shallow sea	
Ocean	——
Tidal range	Small

COLLECTING AND HYDROGRAPHIC FACILITIES	
on shore	——
from boats	Phoronis 10.5 m
Diving (including supply and/or refill of cylinders)	Yes
Collecting service for live material provided by the Station	Yes
Hydrographic data	——

F R A N C E

SPECIAL ORGANISMS AND LOCALITIES THAT ARE A FEATURE OF THE STATION		
name	abundance	season
Extensive brackish water in the Etang de Thau		
Salt pans Extensive sandy shores		
Brackish water fishes Dicentrarchus, Sparus, Anguilla		Autumn Winter
Invertebrates (various)		May-June September

LABORATORIES OPEN TO VISITORS	
General laboratories	Yes
Physiology laboratories	Electrophysiology
Biochemical laboratories	——
Research rooms for visitors	Yes
Wet sorting room	Yes
Experimental aquarium	Large aquarium for aquaculture with tanks 1 to 30m^3
Controlled environment rooms	Yes, temperature
Photographic dark room	Yes
Workshop for general repairs to apparatus	——

F R A N C E

LABORATORY SERVICES	
Electricity - voltage, DC/AC, frequency, phase	220 and 380 V AC 50 Hz also 16 V
Gas	No
Running seawater	Yes
Compressed air (pumps)	Yes
Computer link	No
General glassware	Yes
Chemicals	Yes
Balances (analytical)	No
Balances (torsion)	Yes
Refrigerators	Yes
Ovens	Yes
Microscopes (compound)	Yes
Microscopes (dissecting)	Yes
Electron microscope	No
Centrifuges	Small
Special apparatus	Ultramicrotome, time-lapse cinephotography, Lyophilisator
Library	No information given

F R A N C E

LIVING ACCOMMODATION FOR VISITORS AND THEIR FAMILIES	
At the Station	4 bed rooms
Restaurant	Many in the town
Local Hotels	A wide range available from ungraded to 3 star in the town and nearby
Camp Sites	4 large camps graded 1 - 4 star on the outskirts of the town and nearby
Car parking	Yes

APPROXIMATE CHARGES	
Laboratory	——
Boat	——
Specimens	——
Living accommodation at the Station	5 F a day
Meals at the Station	——
Hotels (range/person/night)	Various
Other	——

F R A N C E

PUBLICATIONS GIVING INFORMATION OF VALUE TO VISITORS

A leaflet giving lodgings, apartments, hotels,
restaurants and camping facilities is available on
request from the Syndicat d'Initiative or L'Office
Municipal du Tourisme

ADDITIONAL INFORMATION SUPPLIED BY THE STATION

None given

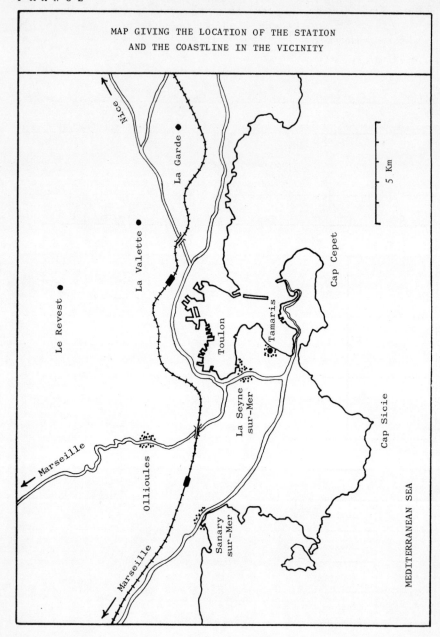

MAP GIVING THE LOCATION OF THE STATION
AND THE COASTLINE IN THE VICINITY

F R A N C E

Name and address of the Station	Institut Michel Pacha Laboratoire maritime de Physiologie 83500 Tamaris-sur-Mer France
	Telephone (94) 94 82 02
Director	Professeur G. Pérès
Affiliation	Université de Lyon

NUMBER OF PERMANENT STAFF	
Scientific	9
Technical	5
Other	2

ROUTES OF ACCESS	
Air	Toulon
Rail	Toulon
Road	Toulon et Marseille
Sea	Toulon, Marseille, Nice

F R A N C E

MAJOR ACCESSIBLE ECOSYSTEMS	
Intertidal	——
Shallow sea	Oui
Ocean	——
Tidal range	——

COLLECTING AND HYDROGRAPHIC FACILITIES	
on shore	Oui
from boats	Oui
Diving (including supply and/or refill of cylinders)	Oui
Collecting service for live material provided by the Station	Oui
Hydrographic data	——

F R A N C E

SPECIAL ORGANISMS AND LOCALITIES THAT ARE A FEATURE OF THE STATION		
name	abundance	season
None quoted		

LABORATORIES OPEN TO VISITORS	
General laboratories	Oui
Physiology laboratories	Oui
Biochemical laboratories	Oui
Research rooms for visitors	Oui
Wet sorting room	Oui
Experimental aquarium	Oui
Controlled environment rooms	Oui
Photographic dark room	Oui
Workshop for general repairs to apparatus	Oui

LABORATORY SERVICES	
Electricity - voltage, DC/AC, frequency, phase	Not given
Gas	Gas supply not given O^2, CO^2, N (in cylinders)
Running seawater	Oui
Compressed air (pumps)	Oui
Computer link	Non
General glassware	Oui
Chemicals	Oui
Balances (analytical)	Oui
Balances (torsion)	Oui
Refrigerators	Oui
Ovens	Oui
Microscopes (compound)	Oui
Microscopes (dissecting)	Oui
Electron microscope	Non
Centrifuges	Oui
Special apparatus	Packard Tricarb
Library	Oui

F R A N C E

LIVING ACCOMMODATION FOR VISITORS AND THEIR FAMILIES	
At the Station	Quelques personnes
Restaurant	Oui
Local Hotels	Oui
Camp Sites	Oui
Car parking	Oui

APPROXIMATE CHARGES	
Laboratory	
Boat	
Specimens	
Living accommodation at the Station	Pas possible vu les circonstances actuelles
Meals at the Station	
Hotels (range/person/night)	
Other	

PUBLICATIONS GIVING INFORMATION OF VALUE TO VISITORS

Revues et ouvrages classiques concernant la Physiologie
et la Biochimie appliquées au domaine maritime

ADDITIONAL INFORMATION SUPPLIED BY THE STATION

None given

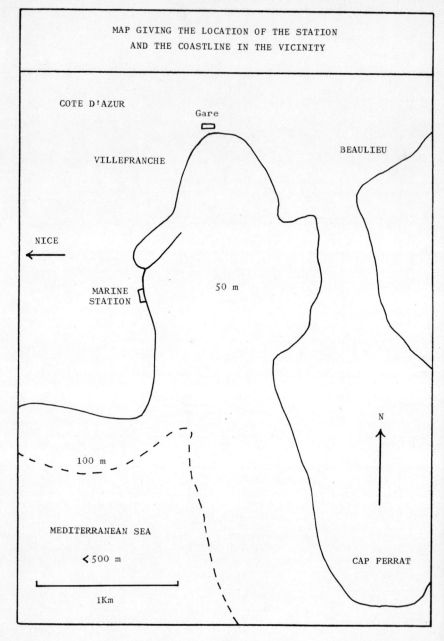

MAP GIVING THE LOCATION OF THE STATION
AND THE COASTLINE IN THE VICINITY

COTE D'AZUR

Gare

VILLEFRANCHE

BEAULIEU

NICE

MARINE
STATION

50 m

N

100 m

MEDITERRANEAN SEA

< 500 m

CAP FERRAT

1Km

F R A N C E

Name and address of the Station	Station Zoologique 06230 Villefranche-sur-Mer France
	Telephone 80-71-65
Director	Professeur P. Bougis, D.Sc.
Affiliation	Université de Paris VI

NUMBER OF PERMANENT STAFF	
Scientific	30
Technical	26
Other	8 marins

ROUTES OF ACCESS	
Air	Nice - de l'aéroport a Villefranche 30 min en automobile
Rail	Nombreux trains de Paris, Marseille ou Gênes. Descendre à Beaulieu (2 km), Nice (7 km) ou Villefranche si possible
Road	L'autoroute Paris-Nice est presque acheree L'autoroute de Gênes s'arrête à 20 km de Villefranche
Sea	Des 'Ferries' relient la Corse à Nice

MAJOR ACCESSIBLE ECOSYSTEMS	
Intertidal	Rocher calcaire. Sable et gravier
Shallow sea	Surface réduite
Ocean	A un quart d'heure pour le bateau du laboratoire
Tidal range	50 cm

COLLECTING AND HYDROGRAPHIC FACILITIES	
on shore	—
from boats	Deux bateaux de 7 m Un bateau de 20 m Filets à plancton de modèles variés Chalut pelagique Isaacs-Kidd Bouteilles de prise d'eau Petites dragues
Diving (including supply and/or refill of cylinders)	Possibilité de gonfler les bouteilles à Nice Bouteilles 'available but no personal equipment on loan'
Collecting service for live material provided by the Station	Pêches de plancton tous les jours
Hydrographic data	Données hebdomadaires sur la temperature et la salinité depuis 1957

SPECIAL ORGANISMS AND LOCALITIES THAT ARE A FEATURE OF THE STATION		
name	abundance	season
Tous les organismes habituels du plancton mediterranéen	variable suivant les espèces	toute l'année
Faune bathypélagique	" "	" "
Oursuis (Paracentrotus, Arbacia)	satisfaisante pour les travaux biologiques	" "

LABORATORIES OPEN TO VISITORS	
General laboratories	Un laboratoire (20 places) pour l'enseignement
Physiology laboratories	Non utilisable pour les visiteurs sans arrangement spécial
Biochemical laboratories	——
Research rooms for visitors	4
Wet sorting room	Dans l'aquarium
Experimental aquarium	Large and small tanks available by arrangement
Controlled environment rooms	Chambres à temperature constante
Photographic dark room	Oui
Workshop for general repairs to apparatus	Oui

LABORATORY SERVICES	
Electricity – voltage, DC/AC, frequency, phase	220 - 380 V AC 50 Hz
Gas	Oui
Running seawater	Non
Compressed air (pumps)	Non
Computer link	Oui
General glassware	Oui
Chemicals	Oui
Balances (analytical)	Oui
Balances (torsion)	Non
Refrigerators	Oui
Ovens	Oui
Microscopes (compound)	Oui
Microscopes (dissecting)	Oui
Electron microscope	Oui après arrangement special
Centrifuges	Oui
Special apparatus	—
Library	Oui

F R A N C E

LIVING ACCOMMODATION FOR VISITORS AND THEIR FAMILIES	
At the Station	Oui - les enfants ne sont pas admis
Restaurant	Oui - une cantine fonctionne à proximité
Local Hotels	Nombreux
Camp Sites	Non
Car parking	Oui

APPROXIMATE CHARGES	
Laboratory	Néant
Boat	Néant
Specimens	Néant
Living accommodation at the Station	5 F
Meals at the Station	6 F 50
Hotels (range/person/night)	20 F - 100 F
Other	Néant

PUBLICATIONS GIVING INFORMATION OF VALUE TO VISITORS

Néant

ADDITIONAL INFORMATION SUPPLIED BY THE STATION

La station s'intèresse principalement au plancton
et à son écologie

NOTES

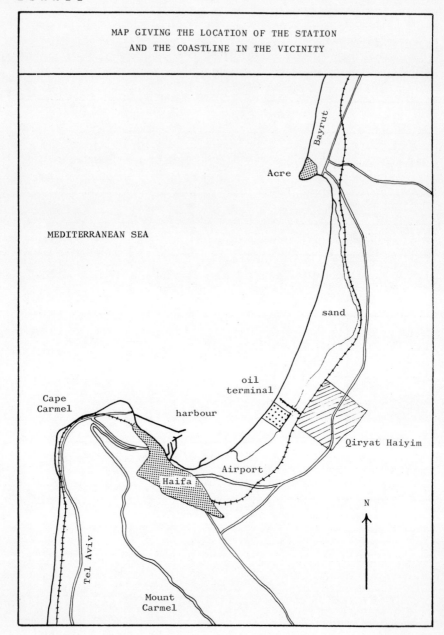

MAP GIVING THE LOCATION OF THE STATION
AND THE COASTLINE IN THE VICINITY

Bayrut

Acre

MEDITERRANEAN SEA

sand

oil
terminal

Cape
Carmel

harbour

Qiryat Haiyim

Airport

Haifa

N

Tel Aviv

Mount
Carmel

I S R A E L

Name and address of the Station	Israel National Institute for Oceanography and Limnology, POB 1793 Haifa Israel
	Telephone 539258/9 642345/6
Director	Admiral (ret) Yohai Ben-Nun
Affiliation	Israel Oceanographic and Limnological Research Co.

NUMBER OF PERMANENT STAFF	
Scientific	4 chemical oceanographers 2 planktologists 5 physical oceanographers 6 marine biologists 1 marine geologist
Technical	8 Technical staff
Other	2 librarians 8 crew members of two boats

ROUTES OF ACCESS	
Air	Lod Airport - 100 km from the Laboratory Elat Airport (Red Sea shore)
Rail	Tel Aviv - 100 km from Haifa Jerusalem - 150 km from Haifa
Road	Tel Aviv (Lod Airport) - Haifa 100 km
Sea	Haifa (from Europe and the Americas) Eilat (from the Far East)

MAJOR ACCESSIBLE ECOSYSTEMS	
Intertidal	Rocks and sand near the laboratory
Shallow sea	Adjacent to the laboratory
Ocean	In the immediate vicinity of Haifa harbour: regular cruises in the Levant sea to Greece, Cyprus etc.
Tidal range	Maximum 50 cm

COLLECTING AND HYDROGRAPHIC FACILITIES	
on shore	Various cars with drivers available, beach seines, jars, etc.
from boats	Catamaran with echosounder, hydrographic winch, Niskin water bottles, reversing thermometers, plankton nets, bottom collecting gear, BT, Decca Navigation System, geological and seismic facilities. Outboard boat.
Diving (including supply and/or refill of cylinders)	Underwater laboratory organized in connection with 'Rambam' Governmental Hospital 2 km away. All diving equipment, decompression chamber etc. available in the laboratory
Collecting service for live material provided by the Station	Scientists collect their own specimens with the aid of the Station's boats and crew
Hydrographic data	Recordings of sea temperature, air, wind etc. since 1946 available at the laboratory

ISRAEL

SPECIAL ORGANISMS AND LOCALITIES THAT ARE A FEATURE OF THE STATION		
name	abundance	season
Common fish and invertebrates		

LABORATORIES OPEN TO VISITORS	
General laboratories	Student laboratory 40 m^2 Guest laboratories 100 m^2
Physiology laboratories	Yes
Biochemical laboratories	Yes
Research rooms for visitors	Yes - 60 m^2 (12 m^2 each)
Wet sorting room	Yes
Experimental aquarium	Yes - 75 m^2
Controlled environment rooms	Yes - 20 m^2
Photographic dark room	Yes - 45 m^2
Workshop for general repairs to apparatus	Electronic and mechanical

LABORATORY SERVICES	
Electricity - voltage, DC/AC, frequency, phase	110; 220; 24; DC/AC 50-60 Hz
Gas	Yes
Running seawater	In biological labs and aquaria
Compressed air (pumps)	Central
Computer link	Many facilities available in the town, IBM research Center at Technion
General glassware	Glass blowing facilities in the town
Chemicals	Large supplies readily available Rare chemicals ordered by air
Balances (analytical)	Many types available in various laboratories
Balances (torsion)	Yes
Refrigerators	Many available in different laboratories
Ovens	Available in chemical laboratory
Microscopes (compound)	Many of various types available
Microscopes (dissecting)	Many types available
Electron microscope	Available in other institutions
Centrifuges	Various available
Special apparatus	Decompression tank, autoanalyzer, gas-chromatograph, spectrophotometers
Library	Excellent; about 1400 journals and periodicals covering marine biology and oceanography, fisheries. Many holdings of important journals complete.

ISRAEL

LIVING ACCOMMODATION FOR VISITORS AND THEIR FAMILIES	
At the Station	Not available
Restaurant	Meals available in cafeteria
Local Hotels	Many with good locations ranging from dormitory to luxury standards. Dormitories in nearby monasteries available.
Camp Sites	Not available
Car parking	At the laboratory

APPROXIMATE CHARGES	
Laboratory	Charges according to the general interest of the Institute in the research
Boat	Free of charge for a limited time
Specimens	Free of charge
Living accommodation at the Station	⎯
Meals at the Station	Approximately $1.0 per meal
Hotels (range/person/night)	US $6 - US $35
Other	Nil

PUBLICATIONS GIVING INFORMATION OF VALUE TO VISITORS

On Haifa town by Government Tourist Information Office;
also available from the Laboratory.

'Bulletin of the Sea Fisheries Research Station'
(discontinued) and the Israel Journal of Zoology, contain
much information on the biota of the area.

ADDITIONAL INFORMATION SUPPLIED BY THE STATION

Main fields of interest of the Institute:

        Marine pollution
        Aquaculture (marine)
        Physical oceanography of the shores
        Physical and chemical oceanography of the
           Levant Basin and the Gulf of Elat
        Fisheries
        Benthos
        Plankton
        Oil prospecting in the sea
        Marine geology (oil prospecting and mineral
           resources)
        Documentation in oceanography
        Man under the sea

The laboratory is built on a rock on the beach and
the walls are washed by the sea.

NOTES

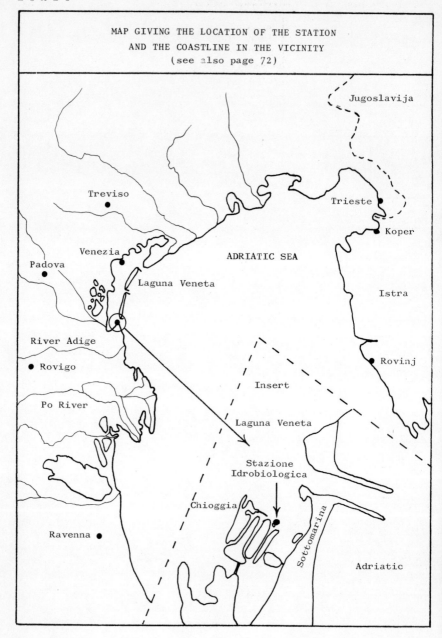

MAP GIVING THE LOCATION OF THE STATION
AND THE COASTLINE IN THE VICINITY
(see also page 72)

Jugoslavija

Treviso

Trieste

Koper

Venezia

ADRIATIC SEA

Padova

Laguna Veneta

River Adige

Istra

Rovigo

Rovinj

Insert

Po River

Laguna Veneta

Stazione
Idrobiologica

Chioggia

Sottomarina

Ravenna

Adriatic

I T A L Y

Name and address of the Station	Stazione Idrobiologica di Chioggia (a division of the Università di Padova) C.P. 101 30015 Chioggia, Italy
	Telephone (041) 400051
Director	Prof. Armando Sabbadin
Affiliation	Università di Padova (address Director c/o) Istituto di Biologia Animale Via Loredan 10, 35100 Padova

NUMBER OF PERMANENT STAFF	
Scientific	Varying number of part-time researchers from the Università di Padova and other institutions
Technical	1 Technician (laboratory + photographic) 1 Boatman 1 Assistant
Other	A bilingual research associate will usually be available to assist foreign visitors with personal logistics and technical problems concerning boats and equipment

ROUTES OF ACCESS	
Air	International flights to Milano or Roma and thence to Venezia (some international flights direct to Venezia)
Rail	Many lines direct to Venezia and thence by bus to Chioggia Rail service from Rovigo to Chioggia but not recommended
Road	Bus service direct from Padova (SIAMIC) and Venezia (leaves from Piazzale Roma near railway station)
Sea	24 h boat service (Acnil line 11) from Venezia; Chioggia and Venezia have international shipping ports

I T A L Y

MAJOR ACCESSIBLE ECOSYSTEMS	
Intertidal	Local beaches, limited tidal flats, abundant man-made vertical surfaces, port entrance jetties
Shallow sea	Inshore waters of the Adriatic
Ocean	——
Tidal range	Mean range about 1 m

COLLECTING AND HYDROGRAPHIC FACILITIES	
on shore	Laboratory has no vehicular transportation but many areas of interest are easily accessible by foot or with public facilities
from boats	Three small boats for use in the lagoon, one of these is capable of light dredging and netting and handling of light hydrographic and other sampling gear; a variety of standard water and bottom sampling gear is available
Diving (including supply and/or refill of cylinders)	Compressor for refilling diving cylinders is available
Collecting service for live material provided by the Station	Limited collecting within the lagoon by the laboratory staff; arrangements can be made with local fishermen for collection in the adjacent areas of the Adriatic
Hydrographic data	Salinity, temperature, DO_2, chlorophyll, turbidity and nutrient data collected daily at the laboratory and periodically in other lagoon areas; limited microbiological data available

66

SPECIAL ORGANISMS AND LOCALITIES THAT ARE A FEATURE OF THE STATION

The Laguna Veneta and Adriatic Sea adjacent to Chioggia have a rich and well documented invertebrate fauna. Chioggia is the major fishing port in the upper Adriatic and a wide variety of species are to be found in the wholesale market. Seasonal abundance varies with species, but fishing activity is sustained throughout most of the year.

The lower lagoon is the site of an intensive mussel culturing industry (Mytilus) with large areas of hanging culture parks.

Brackish water finfish culture is also practised on the periphery of the lagoon and in the Po River delta a short distance south of the city.

LABORATORIES OPEN TO VISITORS	
General laboratories	Chemistry 14 m^2; Microbiol. 9 m^2 Histology 10 m^2; general 30 m^2
Physiology laboratories	None designated specifically for these purposes but one large teaching laboratory can be used when available plus sharing of space in other laboratories
Biochemical laboratories	
Research rooms for visitors	
Wet sorting room	Large wet table and holding tanks
Experimental aquarium	Large outdoor tanks 12 & 8m^3 plus several smaller aquaria
Controlled environment rooms	One room with air and water at 18.5°C
Photographic dark room	Moderately equipped with standard apparatus
Workshop for general repairs to apparatus	Standard repair facilities plus commercial machine shop nearby

LABORATORY SERVICES	
Electricity - voltage, DC/AC, frequency, phase	125/220 V AC 50 Hz single phase 220 V AC 50 Hz 3 phase
Gas	In most laboratories
Running seawater	Two independent systems, one with temperature control
Compressed air (pumps)	Two compressors and independent systems
Computer link	None
General glassware	A moderate stock of standard glassware for chemistry and microbiology
Chemicals	Stock of standard reagents for analysis and media preparation
Balances (analytical)	Three available
Balances	Several trip balances for general weighing and solution preparation
Refrigerators	Three
Ovens	One drying oven and two paraffin ovens
Microscopes (compound)	One binocular and five monocular
Microscopes (dissecting)	Three
Electron microscope	None
Centrifuges	One with capacity 1 litre at 6000 rpm One low speed capacity ca. 30 ml
Special apparatus	Standard equipment for histology; 3 incubators 20°, 37° and 45° and two waterbaths for microbiology; small autoclave; spectrophotometer (B & L mod. 70); pH meter, YSI oxygen meter
Library	Small collection of reference texts Miscellaneous reprint collection Current major marine journals

I T A L Y

LIVING ACCOMMODATION FOR VISITORS AND THEIR FAMILIES	
At the Station	Dormitory facilities for up to five persons Small kitchen at disposal of guests
Restaurant	Many restaurants within walking distance of the laboratory
Local Hotels	The beach area (Sottomarina) is a large seaside resort with many hotels, boarding houses and apartments available on a seasonal basis: Reservation necessary during the summer months
Camp Sites	Many commercial campsites including facilities for trailers are open during the summer season
Car parking	Street parking (no charge) is usually available in the vicinity of the laboratory which is located on a small island accessible only on foot or by boat

APPROXIMATE CHARGES	
Laboratory	There are no fixed charges for temporary or occasional use of the facilities of the Stazione. However, overtime for the services of the technical staff, extra- ordinary expenses associated with excessive boat use, consumption of large quantities or costly reagents etc. would be the responsibility of the visitor
Boat	
Specimens	
Living accommodation at the Station	
Meals at the Station	
Hotels (range/person/night)	——
Other	——

PUBLICATIONS GIVING INFORMATION OF VALUE TO VISITORS

Faganelli, A. 1954. Il trofismo della Laguna Veneta e la
vivificazione marina. I: Ricerche idrografiche.
Arch. Oceanogr. Limnol. vol. 9

Mozzi, C. 1959. Osservazione sull'andamento della temperatura e
della salinità delle acque lagunari di Chioggia in rapporto
alle fasi lunari durante il 1957.
Atti Ist. Ven. Sc. Lett. Arti 94 (II)

Marcuzzi, G. Le collezioni dell'ex istituto di biologia marina
di Rovigno conservate presso la Stazione Idrobiologica di
Chioggia.
Atti Mem. Accad. Patavina Sc. Lett. Arti 84 (II): 169-219, 1972

Brunetti, R. and Canzonier, W.J. 1973. Physico-chemical
observations on the waters of the southern basin of the
Laguna Veneta from 1971 to 1973.
Atti Ist. Veneto Sc. Lett. Arti. 131: 503-523.

ADDITIONAL INFORMATION SUPPLIED BY THE STATION

The Stazione Idrobiologica is custodian of a small museum
collection of animals collected principally in the Upper
Adriatic from 1880 to 1934. Some of these are type
specimens. See the publication of Marcuzzi listed above.

NOTES

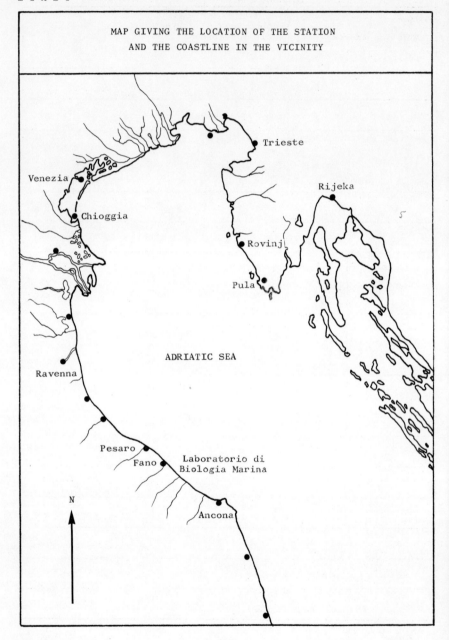

MAP GIVING THE LOCATION OF THE STATION
AND THE COASTLINE IN THE VICINITY

Trieste

Venezia

Chioggia

Rijeka

Rovinj

Pula

ADRIATIC SEA

Ravenna

Pesaro

Fano

Laboratorio di
Biologia Marina

N

Ancona

ITALY

Name and address of the Station	Laboratorio di Biologia marina e Pesca Viale Adriatico 52 61032 Fano Italy
	Telephone 83689
Director	Prof. Andrea Scaccini
Affiliation	University of Bologna

NUMBER OF PERMANENT STAFF	
Scientific	4
Technical	2
Other	2

ROUTES OF ACCESS	
Air	Ancona airport 40 Km Rimini airport 45 Km
Rail	Bologna-Ancona line: Fano station
Road	Strada Flaminia: Roma-Fano Autostrada Bologna-Ancona: Fano exit
Sea	Port of Ancona 50 Km Fishing port at Fano

I T A L Y

MAJOR ACCESSIBLE ECOSYSTEMS	
Intertidal	Negligible
Shallow sea	Yes
Ocean	——
Tidal range	Negligible

COLLECTING AND HYDROGRAPHIC FACILITIES	
on shore	——
from boats	Materials obtainable from fishing boats on request
Diving (including supply and/or refill of cylinders)	——
Collecting service for live material provided by the Station	——
Hydrographic data	Main hydrographic data available at the Station

74

I T A L Y

SPECIAL ORGANISMS AND LOCALITIES THAT ARE A FEATURE OF THE STATION		
name	abundance	season
None quoted		

LABORATORIES OPEN TO VISITORS	
General laboratories	Yes
Physiology laboratories	——
Biochemical laboratories	——
Research rooms for visitors	——
Wet sorting room	——
Experimental aquarium	Yes
Controlled environment rooms	——
Photographic dark room	——
Workshop for general repairs to apparatus	——

LABORATORY SERVICES	
Electricity - voltage, DC/AC, frequency, phase	220 V AC
Gas	Yes
Running seawater	Yes
Compressed air (pumps)	Yes
Computer link	No
General glassware	Yes
Chemicals	Yes
Balances (analytical)	Yes
Balances (torsion)	Yes
Refrigerators	Yes
Ovens	Yes
Microscopes (compound)	Yes
Microscopes (dissecting)	Yes
Electron microscope	No
Centrifuges	Yes
Special apparatus	——
Library	Yes

I T A L Y

LIVING ACCOMMODATION FOR VISITORS AND THEIR FAMILIES	
At the Station	No
Restaurant	Yes
Local Hotels	Yes
Camp Sites	Yes
Car parking	Yes

APPROXIMATE CHARGES	
Laboratory	By arrangement
Boat	By arrangement
Specimens	By arrangement
Living accommodation at the Station	—
Meals at the Station	—
Hotels (range/person/night)	Full board Lit. 5000 - 15000
Other	—

PUBLICATIONS GIVING INFORMATION OF VALUE TO VISITORS

Information on hotels, restaurants and camping etc.
can be obtained from

      Azienda Autonoma di Soggiorno

      Viale C. Colombo

      61032 Fano

ADDITIONAL INFORMATION SUPPLIED BY THE STATION

As space is limited, permission to visit the Station
must be obtained from the Director. Applications
must state the exact nature of the research to be
undertaken, the facilities, space and materials
required, and the proposed length of stay. Team-
based research of common interest is given preference.

MAP GIVING THE LOCATION OF THE STATION
AND THE COASTLINE IN THE VICINITY

ADRIATIC SEA

Lab. Sfrutt.
Biol. Lagune

Lagoons

Lesina

Testa del Gargano

Pescara

Promontorio
del Gargano

S. Severo

Manfredonia

Golfo di
Manfredonia

Foggia

Cerignola

Barletta

Molfetta

Bari

Napoli

N

ITALY

Name and address of the Station	Laboratorio per lo Sfruttamento Biologico delle Lagune CNR Via Fraccacreta, 1 - 71010 Lesina/FG Italy
	Telephone 0882/91166 Cables Lasbla
Director	Febo Lumare
Affiliation	Consiglio Nazionale delle Ricerche

NUMBER OF PERMANENT STAFF	
Scientific	7
Technical	11
Other	6

ROUTES OF ACCESS	
Air	By Itavia from Rome (airport 'Gino Lisa' in Foggia)
Rail	Stations at Foggia, San Severo or Termoli
Road	Toll Highway A 14; Highway S.S. 16
Sea	——

I T A L Y

MAJOR ACCESSIBLE ECOSYSTEMS	
Intertidal	——
Shallow sea	Lagoon
Ocean	——
Tidal range	About 30 cm

COLLECTING AND HYDROGRAPHIC FACILITIES	
on shore	Yes
from boats	Yes
Diving (including supply and/or refill of cylinders)	——
Collecting service for live material provided by the Station	——
Hydrographic data	Available for the Lagoon

I T A L Y

SPECIAL ORGANISMS AND LOCALITIES THAT ARE A FEATURE OF THE STATION		
name	abundance	season
Lagoon system No organisms specified		

LABORATORIES OPEN TO VISITORS	
General laboratories	Yes
Physiology laboratories	——
Biochemical laboratories	——
Research rooms for visitors	——
Wet sorting room	——
Experimental aquarium	——
Controlled environment rooms	Yes
Photographic dark room	——
Workshop for general repairs to apparatus	——

I T A L Y

LABORATORY SERVICES	
Electricity - voltage, DC/AC, frequency, phase	Yes
Gas	Yes
Running seawater	Yes
Compressed air (pumps)	Yes
Computer link	——
General glassware	Yes
Chemicals	Yes
Balances (analytical)	Yes
Balances (torsion)	Yes
Refrigerators	Yes
Ovens	Yes
Microscopes (compound)	Yes
Microscopes (dissecting)	Yes
Electron microscope	——
Centrifuges	Yes
Special apparatus	Spectrophotometer Filter press Thermostatic rooms
Library	Yes

I T A L Y

LIVING ACCOMMODATION FOR VISITORS AND THEIR FAMILIES	
At the Station	1 room with 2 beds and a bathroom
Restaurant	Yes
Local Hotels	Yes
Camp Sites	—
Car parking	Yes

APPROXIMATE CHARGES	
Laboratory	—
Boat	—
Specimens	—
Living accommodation at the Station	—
Meals at the Station	—
Hotels (range/person/night)	$10
Other	—

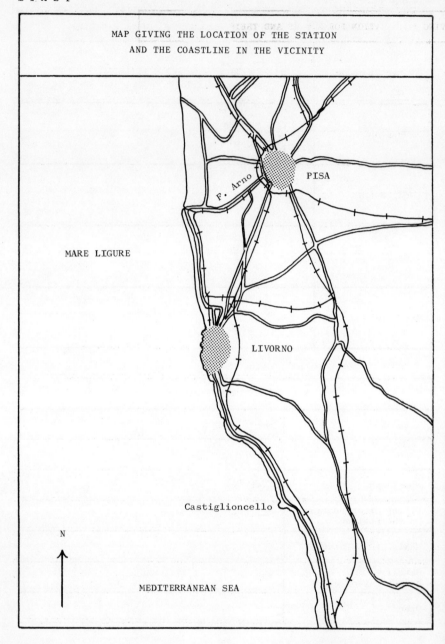

MAP GIVING THE LOCATION OF THE STATION
AND THE COASTLINE IN THE VICINITY

MARE LIGURE

F. Arno

PISA

LIVORNO

Castiglioncello

N

MEDITERRANEAN SEA

I T A L Y

Name and address of the Station	Centro Interuniversitario di Biologia Marina Piazzale Mascagni 1 Livorno Italy
	Telephone 805504
Director	Prof. Paolo Tongiorgi
Affiliation	Universities of Pisa, Firenze, Siena, Bologna, Modena and Torino

NUMBER OF PERMANENT STAFF	
Scientific	One (Prof. Mauro Sordi)
Technical	None
Other	One

ROUTES OF ACCESS	
Air	Pisa: San Giusto Airport (20 Km)
Rail	Pisa (20 Km) Rome (330 Km)
Road	Pisa (20 Km)
Sea	Porto di Livorno (Leghorn)

I T A L Y

MAJOR ACCESSIBLE ECOSYSTEMS	
Intertidal	Yes
Shallow sea	Yes
Ocean	——
Tidal range	——

COLLECTING AND HYDROGRAPHIC FACILITIES	
on shore	Yes
from boats	Yes
Diving (including supply and/or refill of cylinders)	Refill of cylinders only
Collecting service for live material provided by the Station	Yes
Hydrographic data	——

SPECIAL ORGANISMS AND LOCALITIES THAT ARE A FEATURE OF THE STATION		
name	abundance	season
Urchins: Paracentrotus, Arbacia, Spharechinus	plentiful	the whole year
Sea stars: Coscinasterias, Marthasterias	"	but especially in the spring
Holothurians	"	and the summer
Mollusca: Octopus, Aplysia	"	
Crustacea: Carcinus, Pachygrapsus, Portunus	"	
Anemones: Actinia equina, Anemonia sulcata	"	
and many others		

LABORATORIES OPEN TO VISITORS	
General laboratories	Yes
Physiology laboratories	—
Biochemical laboratories	—
Research rooms for visitors	Two or three
Wet sorting room	One
Experimental aquarium	Yes
Controlled environment rooms	—
Photographic dark room	Yes
Workshop for general repairs to apparatus	—

ITALY

LABORATORY SERVICES	
Electricity - voltage, DC/AC, frequency, phase	220 V AC
Gas	Yes
Running seawater	Yes
Compressed air (pumps)	Yes
Computer link	No
General glassware	Yes
Chemicals	Few
Balances (analytical)	Yes
Balances (torsion)	No
Refrigerators	Yes
Ovens	Yes
Microscopes (compound)	Yes
Microscopes (dissecting)	Yes
Electron microscope	No
Centrifuges	Yes
Special apparatus	Spectrophotometer, pH-meter
Library	Yes: a small one

I T A L Y

LIVING ACCOMMODATION FOR VISITORS AND THEIR FAMILIES	
At the Station	None
Restaurant	In the city
Local Hotels	In the city
Camp Sites	Yes: near the city
Car parking	Yes

APPROXIMATE CHARGES	
Laboratory	$10 per day
Boat	$30 per day
Specimens	No
Living accommodation at the Station	No
Meals at the Station	No
Hotels (range/person/night)	No
Other	No

ITALY

MAP GIVING THE LOCATION OF THE STATION
AND THE COASTLINE IN THE VICINITY

Vesuvio

Torre del Greco

Castellammare di Stabla

Golfo di Salerno

Sorrento

Napoli

Stazione Zoologica

Golfo di Napoli

Capri

Pozzuoli

Procida

Golfo di Baeta

Ischia

MEDITERRANEAN SEA

N

ITALY

Name and address of the Station	Stazione Zoologica di Napoli Villa Comunale I - 80121 Napoli Italy
	Telephone 08 1-406.222 Cables Aquarium
Director	Dr. Alessandro Barlaam (interim)
Affiliation	Ministero della Pubblica Istruzione Roma

NUMBER OF PERMANENT STAFF	
Scientific	12
Technical	53
Other	25

ROUTES OF ACCESS	
Air	Naples - Capodichino Airport
Rail	Naples - Central railway station
Road	Autostrada del Sole
Sea	Port of Naples

I T A L Y

MAJOR ACCESSIBLE ECOSYSTEMS	
Intertidal	Yes
Shallow sea	Yes
Ocean	29 Km from Station
Tidal range	ca. 40 cm

COLLECTING AND HYDROGRAPHIC FACILITIES	
on shore	Two trucks
from boats	R. Dohrn (14.5 m); Raffaele (10 m); Lo Bianco (7.50 m); Antonietta (5.40 m); San Gennaro (4.62 m) plus other small boats Echosounder, winch with max. cable length 2000 m (6 mm $\emptyset$), dredges, sledges, grabs, plankton nets, Nansen and van Dorn bottles, etc.
Diving (including supply and/or refill of cylinders)	Three professional divers at disposal for collecting. Arrangements can be made for visitors to dive (insurance and diving ability certificate required). Cylinders supplied and refilled
Collecting service for live material provided by the Station	Boat and shore collecting to meet visitors' requirements
Hydrographic data	Available but sparse

ITALY

SPECIAL ORGANISMS AND LOCALITIES THAT ARE A FEATURE OF THE STATION		
name	abundance	season
Octopus	very common	spring - summer
Sepia	very common	September - May
Loligo	very common	September - May
Paracentrotus	very common	all year round
Arbacia	very common	all year round
Psammechinus	common	all year round
Amphioxus	rare	all year round
Scyliorhinus	common	September - May
Phallusia	rare	all year round
Torpedo	common	September - May
Antedon	common	all year round
Acetabularia	very common	summer
Halimeda	very common	all year round

LABORATORIES OPEN TO VISITORS	
General laboratories	Yes
Physiology laboratories	Yes
Biochemical laboratories	Yes
Research rooms for visitors	Yes
Wet sorting room	Yes
Experimental aquarium	Yes
Controlled environment rooms	Yes (light and temperature)
Photographic dark room	Yes
Workshop for general repairs to apparatus	Yes

ITALY

LABORATORY SERVICES	
Electricity - voltage, DC/AC, frequency, phase	Yes but details not given
Gas	Yes
Running seawater	Yes
Compressed air (pumps)	No
Computer link	Yes
General glassware	Yes
Chemicals	Yes
Balances (analytical)	Yes
Balances (torsion)	Yes
Refrigerators	Yes
Ovens	Yes
Microscopes (compound)	Yes
Microscopes (dissecting)	Yes
Electron microscope	Yes (Philips EM 200)
Centrifuges	Yes
Special apparatus	2 liquid scintillation counters Technicon Autoanalyzer Sterile box for tissue culture Amino acid analyzer
Library	Rebuilt 1957: 80,000 volumes 700 current periodicals Shelf space: 6,300 m

96

I T A L Y

LIVING ACCOMMODATION FOR VISITORS AND THEIR FAMILIES	
At the Station	None available
Restaurant	Yes (lunch only)
Local Hotels	Yes, various grades nearby
Camp Sites	Yes, outside Naples
Car parking	Yes

APPROXIMATE CHARGES	
Laboratory	Research table system covers all facilities including common chemicals, glassware, etc. Table fee: 4,000,000 Italian Lire per year
Boat	
Specimens	
Living accommodation at the Station	____
Meals at the Station	L. 500
Hotels (range/person/night)	1st class L.6,000; 2nd class L.3,000 person/day
Other	Private accommodation 50,000/month

PUBLICATIONS GIVING INFORMATION OF VALUE TO VISITORS

The following are available:

Activity Report of the Naples Zoological Station

Information for guests

Statutes of the Station

ADDITIONAL INFORMATION SUPPLIED BY THE STATION

None given

NOTES

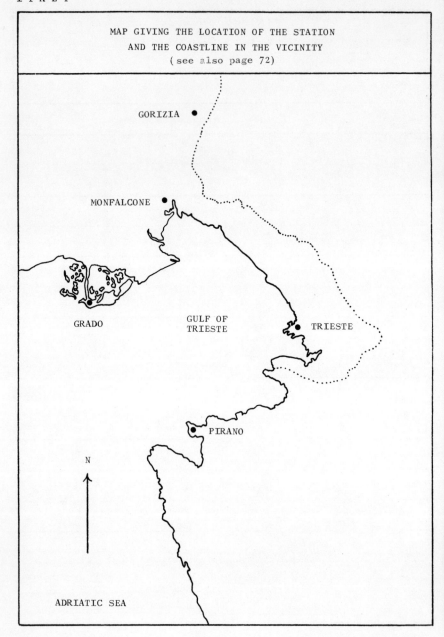

MAP GIVING THE LOCATION OF THE STATION
AND THE COASTLINE IN THE VICINITY
(see also page 72)

GORIZIA

MONFALCONE

GRADO

GULF OF
TRIESTE

TRIESTE

PIRANO

N

ADRIATIC SEA

ITALY

Name and address of the Station	Istituto Sperimentale Talassografico 'F. Vercelli' Viale R. Gessi no2 I-34123 Trieste Italy
	Telephone 35485
Director	Prof. dott. Leopoldo Trotti
Affiliation	Ministero dell'Agricoltura e delle Foreste I - 00100 Rome

NUMBER OF PERMANENT STAFF	
Scientific	4
Technical	2
Other	21

ROUTES OF ACCESS	
Air	Milano international airport with connections to Venice and Ronchi
Rail	Trieste
Road	Trieste
Sea	Trieste

I T A L Y

MAJOR ACCESSIBLE ECOSYSTEMS	
Intertidal	——
Shallow sea	Yes
Ocean	——
Tidal range	About 1 m

COLLECTING AND HYDROGRAPHIC FACILITIES	
on shore	Tide gauges
from boats	——
Diving (including supply and/or refill of cylinders)	——
Collecting service for live material provided by the Station	1 car and 1 lorry
Hydrographic data	Data available for central Mediterranean

I T A L Y

SPECIAL ORGANISMS AND LOCALITIES THAT ARE A FEATURE OF THE STATION		
name	abundance	season
None stated		

LABORATORIES OPEN TO VISITORS	
General laboratories	Yes
Physiology laboratories	—
Biochemical laboratories	Yes
Research rooms for visitors	Yes
Wet sorting room	—
Experimental aquarium	—
Controlled environment rooms	—
Photographic dark room	Yes
Workshop for general repairs to apparatus	Yes

LABORATORY SERVICES	
Electricity – voltage, DC/AC, frequency, phase	125 and 220 V AC 50 Hz single and 3 phase
Gas	Yes
Running seawater	——
Compressed air (pumps)	Yes
Computer link	No, but computer H.P. Packard 9821 A
General glassware	Yes
Chemicals	Yes
Balances (analytical)	Yes
Balances (torsion)	——
Refrigerators	Yes
Ovens	Yes
Microscopes (compound)	Yes
Microscopes (dissecting)	——
Electron microscope	——
Centrifuges	Yes
Special apparatus	Apparatus for general geological research Spectrophotometer, printing facilities on Varitype and reprint copiers, Rank Xerox equipment
Library	650 periodicals; 2000 books; complete set of maps for central Mediterranean (Italian and French)

I T A L Y

LIVING ACCOMMODATION FOR VISITORS AND THEIR FAMILIES	
At the Station	Guest room and bathroom for one
Restaurant	None
Local Hotels	Various
Camp Sites	——
Car parking	Yes

APPROXIMATE CHARGES	
Laboratory	Free
Boat	——
Specimens	——
Living accommodation at the Station	Free
Meals at the Station	——
Hotels (range/person/night)	1st class Lit. 12.000 per night 2nd class Lit. 7.000 per night
Other	——

I T A L Y

PUBLICATIONS GIVING INFORMATION OF VALUE TO VISITORS
None given

ADDITIONAL INFORMATION SUPPLIED BY THE STATION
Annuario and various publications printed at the Institute

MAP GIVING THE LOCATION OF THE STATION
AND THE COASTLINE IN THE VICINITY
(see also page 72)

1   Institute
2   Railway Station and Piazzale Roma
3   Car Parking (Tronchetto)
4   Camping Site on the Lido
5   Airport

VENICE

N

CHIOGGIA

GULF OF VENICE

ADRIATIC SEA

0   10   20 Km

RIVER PO

Name and address of the Station	Istituto di Biologia del Mare Riva Sette Martiri, 1364/A 30122 Venezia Italy	
	Telephone 041/707622	Cables Coricerche Venezia
Director	Professor Bruno Battaglia	
Affiliation	Consiglio Nazionale delle Ricerche (National Research Council)	

NUMBER OF PERMANENT STAFF	
Scientific	1 Oceanographer 1 Chemist 1 Phytoplanktologist 2 Geologists 1 Zooplanktologist 3 Geneticists 1 Microbiologist 1 Ecologist (fouling)
Technical	16 Technical staff
Other	Boat service: 6 (including crew of research vessel) Administrative service: 6 (including 1 Librarian)

ROUTES OF ACCESS	
Air	Venice Airport (bus service to Piazzale Roma, then by boat to the Giardini Biennale, altogether about $1\frac{1}{2}$ hrs.
Rail	Venice Railway Station, then by boat (45 min) to the Giardini Biennale
Road	Autostrada to within 10 Km, then normal road to Piazzale Roma
Sea	Several lines to the Port of Venice

I T A L Y

MAJOR ACCESSIBLE ECOSYSTEMS	
Intertidal	Beaches (outside the Lagoon of Venice), intertidal flats and salt marshes (within the Lagoon)
Shallow sea	Northern Adriatic (depth range: 0 - 80 m)
Ocean	—
Tidal range	About 1 m on average within the Lagoon, atypical semi-diurnal pattern

COLLECTING AND HYDROGRAPHIC FACILITIES	
on shore	1 Fiat laboratory car
from boats	24 m boat (70 metric tons, crew of 4) for oceanographic research (not suitable for trawling) 3 motor-launches 2 rubber-boats
Diving (including supply and/or refill of cylinders)	Visitors interested in this facility are requested to contact the Director
Collecting service for live material provided by the Station	No regular service other than providing material for the research programs of the Institute
Hydrographic data	No continuous sampling from fixed stations (except for those established for special research programs of limited duration)

SPECIAL ORGANISMS AND LOCALITIES THAT ARE A FEATURE OF THE STATION		
name	abundance	season
Mediterranean inshore and lagoon fauna Organisms currently studied in the Genetics Section: Tisbe div. sp. Idotea baltica Gammarus sp. Beach rocks and organogenic rocks outside the Lagoon of Venice		

LABORATORIES OPEN TO VISITORS	
General laboratories	—
Physiology laboratories	—
Biochemical laboratories	—
Research rooms for visitors	No special accommodation available until 1974-75
Wet sorting room	—
Experimental aquarium	Some small tanks available
Controlled environment rooms	2 controlled temperature chambers 1 with programmed lighting
Photographic dark room	One small room
Workshop for general repairs to apparatus	Well-equipped machineshop (1 room) with two mechanics

I T A L Y

LABORATORY SERVICES	
Electricity - voltage, DC/AC, frequency, phase	125 & 220 V AC, 50 Hz, single and 3 phase DC by batteries
Gas	Methane
Running seawater	——
Compressed air (pumps)	Partly piped supply plus portable pumps
Computer link	Not available (electronic desk calculator Olivetti Programma available)
General glassware	Limited supply of common items
Chemicals	Limited supply of common reagents
Balances (analytical)	Various types available
Balances (torsion)	
Refrigerators	Down to - 30°C
Ovens	200°C, 1,200°C
Microscopes (compound)	Oil immersion, phase & opaque illumination
Microscopes (dissecting)	A few available (10 - 40x)
Electron microscope	——
Centrifuges	6,000 r.p.m.
Special apparatus	Gaschromatographic equipment, spectrophoto-meter (single beam), Technicon Autoanalyzer, STD Bissett-Berman, Coulter Counter, available to visitors by special arrangement
Library	Leading marine research journals, some complete series, others complete from about 1967; about 25,000 volumes

I T A L Y

LIVING ACCOMMODATION FOR VISITORS AND THEIR FAMILIES	
At the Station	Apartment with two double bedrooms
Restaurant	Various possibilities (all the year)
Local Hotels	Various possibilities (all the year)
Camp Sites	On the Lido about 40 min trip (bus and boat) from the Institute; also at Punta Sabbioni
Car parking	On the island Tronchetto near Piazzale Roma (10 min by bus & 45 min by boat from the Institute)

APPROXIMATE CHARGES	
Laboratory	——
Boat	——
Specimens	——
Living accommodation at the Station	To be established
Meals at the Station	Not available (for families own cooking possible in the guest-apartment)
Hotels (range/person/night)	Minimum between Lit. 4,000 and 6,000 per night per person full board
Other	——

PUBLICATIONS GIVING INFORMATION OF VALUE TO VISITORS

The Institute will send on request a guide to the city, a list of lodgings and other information available at the Tourist Office.

The description of the Institute, "Il Centro nazionale di studi talassografici" by G. Agricola (Consiglio Nazionale delle Ricerche, Roma, 1968), is no longer up to date and is out of print.

The Institute's Journal, "Archivio di Oceanografia e Limnologia", can be used as a source of information on some of the main research programs currently being pursued at the Institute. Reprints of articles published in this journal and of other publications by the staff of the Institute can be sent on request.

ADDITIONAL INFORMATION SUPPLIED BY THE STATION

Main current research subjects (1973):

Physical and biological oceanography of the Northern Adriatic (excluding fish);

Ecological genetics and population dynamics of marine animals;

Ecology of the Lagoon of Venice with special consideration for the aspects of pollution.

All applications to work at the Institute should be addressed to the Director.

A new building is expected to be completed in 1974-75 and will provide more laboratory space for visitors. Enquiries about accommodation should be made to the Director.

NOTES

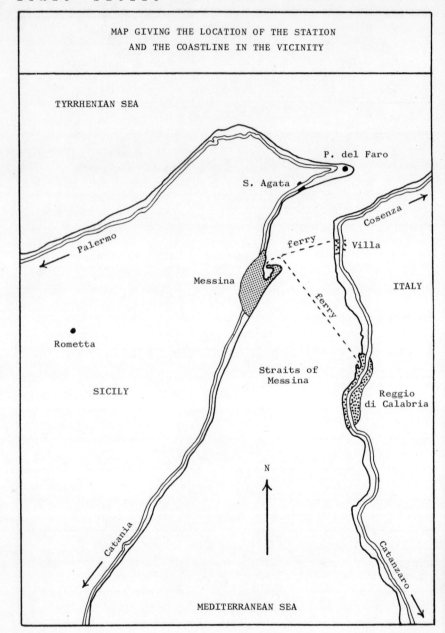

MAP GIVING THE LOCATION OF THE STATION
AND THE COASTLINE IN THE VICINITY

TYRRHENIAN SEA

P. del Faro

S. Agata

Cosenza

Palermo

ferry

Villa

Messina

ferry

ITALY

Rometta

Straits of
Messina

SICILY

Reggio
di Calabria

N

Catania

Catanzaro

MEDITERRANEAN SEA

ITALY - SICILY

Name and address of the Station	Stazione di Biologia Marina dell' Istituto di Zoologia dell' Università di Messina Sicily
	Telephone 090/812721
Director	Prof. Arturo Bolognari
Affiliation	University of Messina Ministero della Pubblica Istruzione

NUMBER OF PERMANENT STAFF	
Scientific	8
Technical	5
Other	5

ROUTES OF ACCESS	
Air	Reggio Calabria or Catania Airport
Rail	Messina Station
Road	Yes
Sea	Yes, Messina harbour

ITALY - SICILY

MAJOR ACCESSIBLE ECOSYSTEMS	
Intertidal	Limited
Shallow sea	Nearby and offshore
Ocean	Deep water offshore
Tidal range	Negligible

COLLECTING AND HYDROGRAPHIC FACILITIES	
on shore	Yes
from boats	Two boats: Colapesce 13 m accommodates 3 researchers. Algesiro 29 m accommodates 8 - 10 researchers
Diving (including supply and/or refill of cylinders)	"Scuola per biologi subacquei" provides courses for biology students in scuba diving
Collecting service for live material provided by the Station	Yes
Hydrographic data	Yes

SPECIAL ORGANISMS AND LOCALITIES THAT ARE A FEATURE OF THE STATION		
name	abundance	season
Straits of Messina and its features		
Abyssal fish stranded on the beach	Considerable	All the year round
Laminariales	"	Spring - Summer
Xiphias gladius	"	" "
Phytoplankton	"	All the year round

LABORATORIES OPEN TO VISITORS	
General laboratories	Yes
Physiology laboratories	——
Biochemical laboratories	——
Research rooms for visitors	——
Wet sorting room	Yes
Experimental aquarium	Yes
Controlled environment rooms	——
Photographic dark room	Yes
Workshop for general repairs to apparatus	Yes

LABORATORY SERVICES	
Electricity - voltage, DC/AC, frequency, phase	220 V AC 50 Hz single phase 380 V AC 50 Hz three phase
Gas	Yes
Running seawater	No
Compressed air (pumps)	Yes
Computer link	Available at Science Faculty
General glassware	Yes
Chemicals	Yes
Balances (analytical)	Yes
Balances (torsion)	Yes
Refrigerators	Yes
Ovens	Yes
Microscopes (compound)	Yes
Microscopes (dissecting)	Yes
Electron microscope	Available at Science Faculty
Centrifuges	Yes
Special apparatus	Flow Geiger-Muller for ^{14}C technique Atomic absorption double beam - Perkin Elmer, Fluorescence and U.V. spectro-photometers
Library	Yes

LIVING ACCOMMODATION FOR VISITORS AND THEIR FAMILIES	
At the Station	None available
Restaurant	Yes
Local Hotels	Yes
Camp Sites	Yes
Car parking	Yes

APPROXIMATE CHARGES	
Laboratory	
Boat	
Specimens	
Living accommodation at the Station	Not quoted
Meals at the Station	
Hotels (range/person/night)	
Other	

ITALY - SICILY

PUBLICATIONS GIVING INFORMATION OF VALUE TO VISITORS

Bolognari, A.   Marine Biological Station - Zoological
    Institute - University of Messina. Report on
    research activities for the years 1967-71.
    Mem. Biol. Mar. Ocean. 1, 1-19, 1971.

ADDITIONAL INFORMATION SUPPLIED BY THE STATION

None

Name and address of the Station	Istituto Sperimentale Talassografico Spianata S. Raineri 6-98100 Messina (Piazza Todaro) Italy (see map on page 116)
	Telephone 48.130
Director	Dr. Antonino Cavaliere
Affiliation	Ministero dell'Agricoltura e delle Foreste

NUMBER OF PERMANENT STAFF	
Scientific	2
Technical	1
Other	6

ROUTES OF ACCESS	
Air	Reggio Calabria or Catania Airport
Rail	Messina
Road	Via S. Raineri, 6
Sea	Messina harbour

MAJOR ACCESSIBLE ECOSYSTEMS	
Intertidal	Yes
Shallow sea	Yes, nearby
Ocean	Deep water offshore
Tidal range	Negligible

COLLECTING AND HYDROGRAPHIC FACILITIES	
on shore	Yes
from boats	Yes
Diving (including supply and/or refill of cylinders)	—
Collecting service for live material provided by the Station	On request
Hydrographic data	Yes

ITALY - SICILY

SPECIAL ORGANISMS AND LOCALITIES THAT ARE A FEATURE OF THE STATION		
name	abundance	season
Stomiatoidea	relatively abundant	Autumn - Winter
Mictofoidea	"	"
Scombroidei	"	Spring - Summer

LABORATORIES OPEN TO VISITORS	
General laboratories	—
Physiology laboratories	—
Biochemical laboratories	Yes
Research rooms for visitors	Yes
Wet sorting room	Yes
Experimental aquarium	Yes
Controlled environment rooms	—
Photographic dark room	Yes
Workshop for general repairs to apparatus	Yes

LABORATORY SERVICES	
Electricity - voltage, DC/AC, frequency, phase	Yes
Gas	Bottle gas
Running seawater	Yes
Compressed air (pumps)	——
Computer link	——
General glassware	Yes
Chemicals	Yes
Balances (analytical)	Yes
Balances (torsion)	——
Refrigerators	Yes
Ovens	Yes
Microscopes (compound)	Yes
Microscopes (dissecting)	——
Electron microscope	——
Centrifuges	Yes
Special apparatus	Current meter, refractometer, spectro-photometer, bathythermograph, bottles for measuring water temperature
Library	Yes

LIVING ACCOMMODATION FOR VISITORS AND THEIR FAMILIES	
At the Station	No
Restaurant	No
Local Hotels	Yes
Camp Sites	Outside the City
Car parking	Yes

APPROXIMATE CHARGES	
Laboratory	By arrangement
Boat	
Specimens	
Living accommodation at the Station	No
Meals at the Station	No
Hotels (range/person/night)	1st class Lit. 10,000; 2nd class Lit.7000; 3rd class Lit. 5000 per person per day
Other	——

ITALY - SICILY

PUBLICATIONS GIVING INFORMATION OF VALUE TO VISITORS

General tourist information may be obtained from:

Ente Provinciale Turismo - Telef. 48140
Azienda Autonoma di Soggiorno e Turismo - Telef. 36494

ADDITIONAL INFORMATION SUPPLIED BY THE STATION

The Institute is situated 3 Km from the City of
Messina overlooking the Straits and the port.
The entrance is adjacent to that of the Marina
Militare.

The Institute's aquarium is in the northern part
of the city in the Villa Mazzini behind the
Palazzo della Prefettura.

NOTES

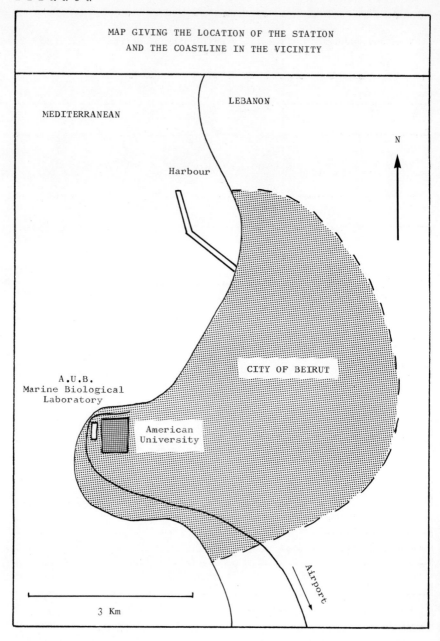

MAP GIVING THE LOCATION OF THE STATION
AND THE COASTLINE IN THE VICINITY

MEDITERRANEAN

LEBANON

N

Harbour

A.U.B.
Marine Biological
Laboratory

CITY OF BEIRUT

American
University

Airport

3 Km

L E B A N O N

Name and address of the Station	Marine Science Program Department of Biology American University of Beirut Beirut Lebanon
	Telephone 340-740 ext.2409
Director	Dr. John T. Hardy
Affiliation	American University of Beirut

NUMBER OF PERMANENT STAFF	
Scientific	1 Marine Botanist 1 Marine Geologist 1 Invertebrate Zoologist 1 Marine Microbiologist 1 Ichthyologist 1 Zooplanktonologist 1 Marine Chemist 1 Phytoplanktonologist
Technical	5 Research assistants
Other	—

ROUTES OF ACCESS	
Air	To Beirut Airport
Rail	From Istanbul
Road	From Istanbul
Sea	World wide connections

L E B A N O N

MAJOR ACCESSIBLE ECOSYSTEMS	
Intertidal	Limestone platforms and sandy beaches next to research laboratory
Shallow sea	Limited
Ocean	Easy access via Port of Beirut (2 Km)
Tidal range	0 - 30 cm

COLLECTING AND HYDROGRAPHIC FACILITIES	
on shore	Cars and carryalls available from University motor pool
from boats	4 m outboard runabout at laboratory Larger vessel sometimes available on charter
Diving (including supply and/or refill of cylinders)	SCUBA equipment and air available at the laboratory
Collecting service for live material provided by the Station	Divers usually available for submarine collecting
Hydrographic data	New oceanographic data buoy presently under construction

SPECIAL ORGANISMS AND LOCALITIES THAT ARE A FEATURE OF THE STATION		
name	abundance	season
Herbarium collections available		

LABORATORIES OPEN TO VISITORS	
General laboratories	Available
Physiology laboratories	——
Biochemical laboratories	——
Research rooms for visitors	Available on limited basis
Wet sorting room	Available
Experimental aquarium	Yes
Controlled environment rooms	Yes
Photographic dark room	Yes
Workshop for general repairs to apparatus	Limited facilities

LEBANON

LABORATORY SERVICES	
Electricity - voltage, DC/AC, frequency, phase	110 V AC, 50 Hz
Gas	Yes
Running seawater	Yes
Compressed air (pumps)	Yes
Computer link	Limited facilities
General glassware	Yes
Chemicals	Yes
Balances (analytical)	Yes
Balances (torsion)	Yes
Refrigerators	Yes
Ovens	Yes
Microscopes (compound)	Yes
Microscopes (dissecting)	Yes
Electron microscope	In Medical School
Centrifuges	Yes
Special apparatus	Wide variety of marine biological and simple oceanographic gear, isotope equipment, etc.
Library	At the American University of Beirut

L E B A N O N

LIVING ACCOMMODATION FOR VISITORS AND THEIR FAMILIES	
At the Station	Possible in summer
Restaurant	Many local
Local Hotels	Many
Camp Sites	None
Car parking	Yes

APPROXIMATE CHARGES	
Laboratory	No charge
Boat	$12 per day
Specimens	No charge
Living accommodation at the Station	Summer apartments — about $160 per month
Meals at the Station	None
Hotels (range/person/night)	Not given
Other	None

PUBLICATIONS GIVING INFORMATION OF VALUE TO VISITORS

Publications will be sent on request

ADDITIONAL INFORMATION SUPPLIED BY THE STATION

Enquiries should be made to

Dr. John Hardy
Co-Chairman
Marine Science Program
American University of Beirut
Beirut
Lebanon

NOTES

MAP GIVING THE LOCATION OF THE STATION
AND THE COASTLINE IN THE VICINITY

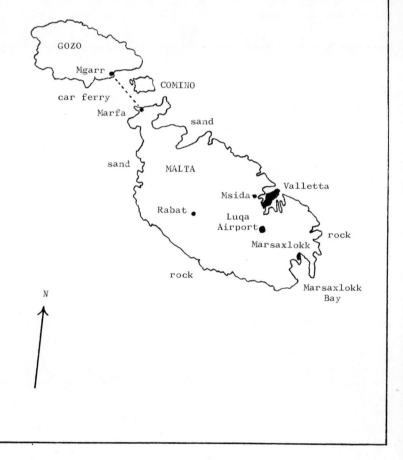

MALTA

Name and address of the Station	International Ocean Institute Royal University of Malta Msida Malta
	Telephone 36451 ext. 294
Director	Sidney J. Holt
Affiliation	An international non-governmental organisation established legally within the Royal University of Malta

NUMBER OF PERMANENT STAFF	
Scientific	One
Technical	One
Other	One

ROUTES OF ACCESS	
Air	From London, Paris, Rome, Catania and Tripoli
Rail	——
Road	——
Sea	From Sicily and S. Italy (car ferries)

MALTA

MAJOR ACCESSIBLE ECOSYSTEMS	
Intertidal	Spray pools
Shallow sea	Limestone reef; rock and sand bottom
Ocean	Deep water close to shore in some directions
Tidal range	Negligible

COLLECTING AND HYDROGRAPHIC FACILITIES	
on shore	Some hydrographic and biological sampling equipment available, mostly such as is used for training programmes for undergraduate students
from boats	
Diving (including supply and/or refill of cylinders)	Easily and cheaply arranged, by rental of equipment and air
Collecting service for live material provided by the Station	——
Hydrographic data	——

MALTA

LIVING ACCOMMODATION FOR VISITORS AND THEIR FAMILIES	
At the Station	None
Restaurant	At the University
Local Hotels	Ample within 1 - 2 Km but space depends on tourist season
Camp Sites	None, except rough camping on shore sites, by permission, for expeditions
Car parking	Plenty

APPROXIMATE CHARGES	
Laboratory	——
Boat	——
Specimens	——
Living accommodation at the Station	——
Meals at the Station	Student canteen prices
Hotels (range/person/night)	Range from very simple and cheap to luxury
Other	——

PUBLICATIONS GIVING INFORMATION OF VALUE TO VISITORS

Malta Tourist Office brochures and many tourist guides
available through Airlines and Travel Agencies

Reports of the International Ocean Institute indicate
its activities

The Institute will answer specific questions on
request

ADDITIONAL INFORMATION SUPPLIED BY THE STATION

Bibliographies and documents of 'Pacem in Moribus'
Conventions and related material on marine resources
and Law of the sea

At present the Station is in its infancy.   Visitors
can be offered office and library space.   There is
a small special library on ocean problems ranging
over the natural and social sciences, politics, law
and international affairs

A new Marine Station is planned in a fort overlooking
Marsaxlokk Bay

MAP GIVING THE LOCATION OF THE STATION
AND THE COASTLINE IN THE VICINITY

R U M A N I A

Name and address of the Station	Institutul Român de Cercetări Marine (Rumanian Institute for Marine Research) Boulevard Lenin n-300 Constantza Rumania
	Telephone 13288
Director	Not given
Affiliation	Rumanian National Council for Science and Technology

NUMBER OF PERMANENT STAFF	
Scientific	More than 100
Technical	
Other	——

ROUTES OF ACCESS	
Air	Constantza airport
Rail	Constantza station
Road	Bucarest-Constantza
Sea	Constantza harbour

RUMANIA

MAJOR ACCESSIBLE ECOSYSTEMS	
Intertidal	None
Shallow sea	Depths down to 2770 m
Ocean	see above
Tidal range	Nil

COLLECTING AND HYDROGRAPHIC FACILITIES	
on shore	Yes
from boats	Yes
Diving (including supply and/or refill of cylinders)	Yes, both facilities
Collecting service for live material provided by the Station	Yes
Hydrographic data	Yes

146

SPECIAL ORGANISMS AND LOCALITIES THAT ARE A FEATURE OF THE STATION		
name	abundance	season
Brackish water organisms Salinity of Black Sea about $19^o/_{oo}$		

LABORATORIES OPEN TO VISITORS	
General laboratories	Yes
Physiology laboratories	Yes
Biochemical laboratories	Yes
Research rooms for visitors	——
Wet sorting room	Yes
Experimental aquarium	Yes
Controlled environment rooms	Yes
Photographic dark room	Yes
Workshop for general repairs to apparatus	Yes

LABORATORY SERVICES	
Electricity - voltage, DC/AC, frequency, phase	220 V AC
Gas	Yes
Running seawater	Yes
Compressed air (pumps)	Yes
Computer link	——
General glassware	Yes
Chemicals	Yes
Balances (analytical)	Yes
Balances (torsion)	Yes
Refrigerators	Yes
Ovens	Yes
Microscopes (compound)	Yes
Microscopes (dissecting)	Yes
Electron microscope	——
Centrifuges	Yes
Special apparatus	——
Library	Yes, more than 5000 volumes

R U M A N I A

LIVING ACCOMMODATION FOR VISITORS AND THEIR FAMILIES	
At the Station	—
Restaurant	—
Local Hotels	In Constantza
Camp Sites	Yes
Car parking	Yes

APPROXIMATE CHARGES	
Laboratory	
Boat	
Specimens	
Living accommodation at the Station	Varies according to season
Meals at the Station	
Hotels (range/person/night)	
Other	

RUMÁNIA

PUBLICATIONS GIVING INFORMATION OF VALUE TO VISITORS

Travaux de l'Inst. Roum. de Recherches Marines
vol. I, II, III. (parus)

ADDITIONAL INFORMATION SUPPLIED BY THE STATION

Further information should be sought from the Director

The section concerned with biological research is at
the Station Agigea 12 Km to the south of Constantza

SPAIN

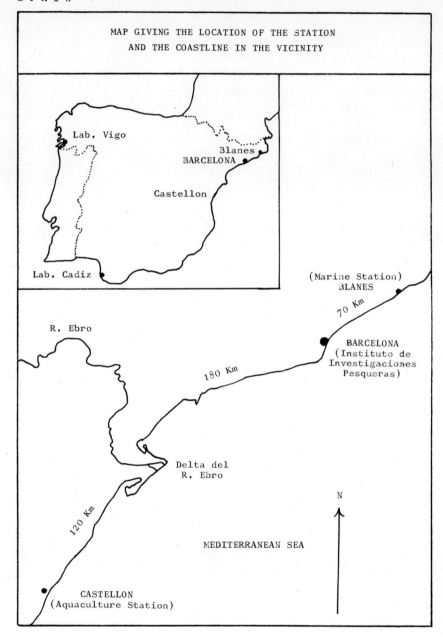

MAP GIVING THE LOCATION OF THE STATION
AND THE COASTLINE IN THE VICINITY

Lab. Vigo

Blanes
BARCELONA

Castellon

Lab. Cadiz

(Marine Station)
BLANES

70 Km

R. Ebro

180 Km

BARCELONA
(Instituto de
Investigaciones
Pesqueras)

Delta del
R. Ebro

N

120 Km

MEDITERRANEAN SEA

CASTELLON
(Aquaculture Station)

SPAIN

Name and address of the Station	Instituto de Investigaciones Pesqueras Paseo Nacional, s/n Barcelona - 3 Spain
	Telephone 319 43 28
Director	Prof. B. Andreu, D.Sc.
Affiliation	Patronato de Investigación Científica y Técnica 'Juan de la Cierva', del Consejo Superior de Investigaciones Científicas

NUMBER OF PERMANENT STAFF	
Scientific	4 marine biologists 2 ecologists 4 zoologists 4 chemists 2 botanists
Technical	9 research assistants
Other	3 boat crew 5 subalterns 5 various

ROUTES OF ACCESS	
Air	Prat airport (7 Km from the town)
Rail	Overnight express from Madrid and Paris
Road	High way from the French border through Junquera
Sea	Shipping lines between America and Europe call at Barcelona

SPAIN

MAJOR ACCESSIBLE ECOSYSTEMS	
Intertidal	Tides hardly perceptible
Shallow sea	Large continental platform. Estuary of the river Ebro (130 Km). Rocky coast at Blanes (70 Km)
Ocean	Accessible by hired boat
Tidal range	20 - 30 cm. Equinoctial 50 cm

COLLECTING AND HYDROGRAPHIC FACILITIES	
on shore	Simple collecting
from boats	Small boat (8 m) but possible to hire fishing boats or to obtain samples from the fishing boats. Plankton nets, midwater nets, reversing bottles, and small drags
Diving (including supply and/or refill of cylinders)	2 sets of diving equipment. Compressed air supplied in the City
Collecting service for live material provided by the Station	Materials can be obtained using the fisheries boats or from the Station's boat
Hydrographic data	Hydrographic information available

SPECIAL ORGANISMS AND LOCALITIES THAT ARE A FEATURE OF THE STATION		
name	abundance	season
Phytoplankton and zooplankton	Moderate	Mainly in spring and autumn
Sepia officinalis	Frequent	Spring
Octopus vulgaris	"	Winter
Eledone cirrhosa	"	May - September
Astropecten aurantiacus	"	All the year
Strongylocentrotus lividus	"	Winter
Aristeus antennatus	"	All the year
Squilla mantis	"	Winter
Spicara maena	"	All the year
Diplodus annularis	"	All the year
Sardina pilchardus	"	All the year
Micromessistius poutassou	"	Winter

LABORATORIES OPEN TO VISITORS	
General laboratories	1 laboratory with 2-3 places
Physiology laboratories	1-2 places for qualified visitors
Biochemical laboratories	Laboratory with 1 place for qualified visitors
Research rooms for visitors	2 single laboratories
Wet sorting room	——
Experimental aquarium	Large and small tanks available
Controlled environment rooms	1 constant temperature room
Photographic dark room	Not open to visitors
Workshop for general repairs to apparatus	Small mechanical and electronic workshop

LABORATORY SERVICES	
Electricity - voltage, DC/AC, frequency, phase	125 and 220 V AC 50 Hz, 3 phase
Gas	Coal gas
Running seawater	In all laboratories
Compressed air (pumps)	Piped compressed air up to 3 atmospheres
Computer link	——
General glassware	——
Chemicals	All kinds
Balances (analytical)	2 (precision)
Balances (torsion)	——
Refrigerators	2 general refrigerators and 2 deep-freezers
Ovens	1000 °C
Microscopes (compound)	Ordinary microscopes and immersion, phase contrast and reverse microscopes
Microscopes (dissecting)	Several types
Electron microscope	——
Centrifuges	Up to 20.000 rpm
Special apparatus	A wide variety of analytical apparatus Autoanalyser Technicon, gas chromatograph, Warburg, spectrophotometer, pH meter, salinometer, autoclave, etc.
Library	400 Marine journals (some since 1949) 2065 volumes

S P A I N

LIVING ACCOMMODATION FOR VISITORS AND THEIR FAMILIES	
At the Station	No bedrooms for visitors
Restaurant	Several restaurants of different classes near the Institute No meals are served at the Station
Local Hotels	Hotel near the Institute available out of the tourist season Comfortable hotels in the old part of the town (consult CSIC)
Camp Sites	In Castelldefels, 5 - 10 Km from the City, good facilities
Car parking	Good parking near the Institute

APPROXIMATE CHARGES	
Laboratory	Usually no charge except for special requirements
Boat	Usually no charge
Specimens	Usually no charge if easily obtained
Living accommodation at the Station	——
Meals at the Station	——
Hotels (range/person/night)	Between 200 and 500 pesetas daily
Other	——

PUBLICATIONS GIVING INFORMATION OF VALUE TO VISITORS

The review of the Institute 'Investigación Pesquera'
gives information on hydrography, primary production,
phyto and zooplankton, and the faunistic and other
activities of the Centre.

Tourist guides of the region may be purchased.

ADDITIONAL INFORMATION SUPPLIED BY THE STATION

The activities of the Institute are oriented mainly
toward research in hydrography, primary production,
bacteriology, planktology, marine biology and the
fisheries of the area from Spain to the Northwest
African coast.   A new method of automatic continuous
analysis of physical, chemical and biological
parameters is used.

The Institute has laboratories in Cadiz and Vigo,
a Station in Blanes and a Pilot Plant for Aquaculture
in Castellon.

SPAIN

Name and address of the Station	Aquarium Apartado, 15 Blanes (Costa Brava) Spain (see map on page 152)
	Telephone 330329
Director	Professor Dr. Manuel Rubió
Affiliation	Instituto Investigaciones Pesqueras (Consejo Superior de Investigaciones Cientificas)

NUMBER OF PERMANENT STAFF	
Scientific	3 zoologists 1 research assistant
Technical	1 technical staff
Other	——

ROUTES OF ACCESS	
Air	Airport Gerona-Costa Brava
Rail	To Blanes (via Cerbère/Massanet-Massanes)
Road	9 Km fast road La Junquera/Barcelona
Sea	——

S P A I N

MAJOR ACCESSIBLE ECOSYSTEMS	
Intertidal	Oceanic sand - sheltered rock (Costa Brava)
Shallow sea	Open sea
Ocean	——
Tidal range	Negligible

COLLECTING AND HYDROGRAPHIC FACILITIES	
on shore	——
from boats	——
Diving (including supply and/or refill of cylinders)	Air àvailable at Blanes harbour
Collecting service for live material provided by the Station	Invertebrates supplied
Hydrographic data	——

SPECIAL ORGANISMS AND LOCALITIES THAT ARE A FEATURE OF THE STATION		
name	abundance	season
Porifera Coelenterata Echinodermata Polychaeta Crustacea Tunicata		

LABORATORIES OPEN TO VISITORS	
General laboratories	2 laboratories for teaching and research
Physiology laboratories	――
Biochemical laboratories	1 laboratory not available to visitors
Research rooms for visitors	1 single; 1 double
Wet sorting room	――
Experimental aquarium	Small tanks
Controlled environment rooms	――
Photographic dark room	――
Workshop for general repairs to apparatus	――

S P A I N

LABORATORY SERVICES	
Electricity - voltage, DC/AC, frequency, phase	200 - 380 V AC 50 Hz
Gas	——
Running seawater	In all laboratories
Compressed air (pumps)	——
Computer link	——
General glassware	——
Chemicals	——
Balances (analytical)	——
Balances (torsion)	——
Refrigerators	——
Ovens	——
Microscopes (compound)	——
Microscopes (dissecting)	——
Electron microscope	——
Centrifuges	——
Special apparatus	——
Library	Main marine journals and books

SPAIN

LIVING ACCOMMODATION FOR VISITORS AND THEIR FAMILIES	
At the Station	A double bedroom
Restaurant	——
Local Hotels	Many available out of tourist season
Camp Sites	There are plenty nearby
Car parking	Yes

APPROXIMATE CHARGES	
Laboratory	——
Boat	——
Specimens	——
Living accommodation at the Station	——
Meals at the Station	——
Hotels (range/person/night)	——
Other	——

PUBLICATIONS GIVING INFORMATION OF VALUE TO VISITORS

Nil

ADDITIONAL INFORMATION SUPPLIED BY THE STATION

This is a small biological station, mainly for
work on species (alive or recently dead) of the
Western Mediterranean, chiefly from deep areas.

Local fishing boats catch at depths down to
800/1000 m.

NOTES

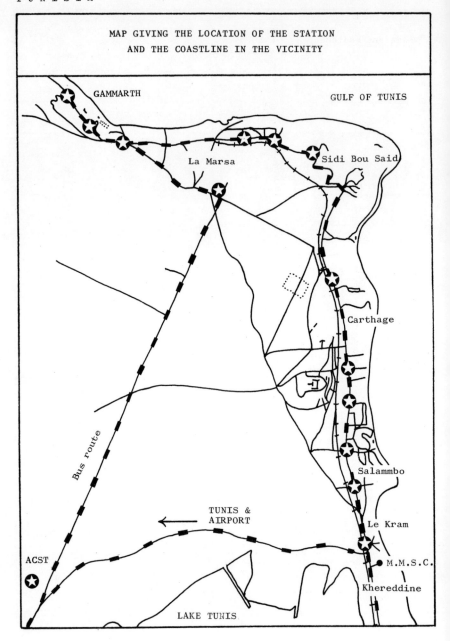

MAP GIVING THE LOCATION OF THE STATION
AND THE COASTLINE IN THE VICINITY

GAMMARTH

GULF OF TUNIS

La Marsa

Sidi Bou Said

Carthage

Bus route

Salammbo

TUNIS &
AIRPORT

Le Kram

ACST

M.M.S.C.

Khereddine

LAKE TUNIS

T U N I S I A

Name and address of the Station	Mediterranean Marine Sorting Center Khereddine Tunisia
	Telephone 276 552 Cables MMSC 276 808 Khereddine
Director	Mr. Ernani G. Menez
Affiliation	Smithsonian Oceanographic Sorting Center Smithsonian Institution Washington, D.C. 20560 U. S. A.

NUMBER OF PERMANENT STAFF	
Scientific	2 Zoologists 2 Botanists 1 Cytogeneticist 4 Research Associates (Biologists)
Technical	20 Technical Staff
Other	5 Administrative Staff 5 General Services

ROUTES OF ACCESS	
Air	To Tunis with a 10 min drive (taxi) to Khereddine
Rail	From Tunis to Khereddine, $\frac{1}{2}$ hr train journey
Road	Highway from Tunis, 13 Km to Khereddine
Sea	Ferries from Marseilles, Genoa and Naples all year round

MAJOR ACCESSIBLE ECOSYSTEMS	
Intertidal	Sandy beaches (100m) – rocky beaches (30Km) – Estuarine mud – Lake of Tunis (hypersaline – muddy)
Shallow sea	Very large continental shelf
Ocean	———
Tidal range	Slight, unpredictable

COLLECTING AND HYDROGRAPHIC FACILITIES	
on shore	1 Peugeot 404 Station Wagon – no driver available. Standard shore collecting gear available (coring tubes, baskets, etc.)
from boats	2 Zodiac (small) with outboard motors
Diving (including supply and/or refill of cylinders)	Compressor available Tanks and regulators must be supplied by individual
Collecting service for live material provided by the Station	Material for study obtained with approval of the Institut Scientifique et Technique d'Océanographie et de Pêche, through MMSC assistance
Hydrographic data	Sea temperature, salinity, winds, etc. are available in the oceanographic sponsoring institution (INSTOP)

T U N I S I A

SPECIAL ORGANISMS AND LOCALITIES THAT ARE A FEATURE OF THE STATION		
name	abundance	season
Common Mediterranean fauna and flora		

LABORATORIES OPEN TO VISITORS	
General laboratories	1 laboratory
Physiology laboratories	Yes
Biochemical laboratories	Yes
Research rooms for visitors	1 double room
Wet sorting room	2 rooms available
Experimental aquarium	Small tanks available and air pumps
Controlled environment rooms	Yes
Photographic dark room	Available
Workshop for general repairs to apparatus	Small repairs

LABORATORY SERVICES	
Electricity - voltage, DC/AC, frequency, phase	110 V and 220 V AC, 50 Hz
Gas	Butane
Running seawater	Not available
Compressed air (pumps)	Available
Computer link	Yes
General glassware	Available
Chemicals	Standard preserving solutions
Balances (analytical)	Mettler available
Balances (torsion)	Yes
Refrigerators	Available
Ovens	Yes (80°C)
Microscopes (compound)	1 available - Phase Contrast - Normaki - photomicrography
Microscopes (dissecting)	Very few available
Electron microscope	——
Centrifuges	——
Special apparatus	Some (enquiry)
Library	Some marine journals available

LIVING ACCOMMODATION FOR VISITORS AND THEIR FAMILIES	
At the Station	1 double bedroom
Restaurant	Breakfast and midday meal could be arranged Normally no meals available at the Station
Local Hotels	Many available out of tourist season (May - Oct)
Camp Sites	None
Car parking	Available

APPROXIMATE CHARGES	
Laboratory	
Boat	No charge for approved usage
Specimens	
Living accommodation at the Station	
Meals at the Station	About US $2.00 per day - 1 meal + breakfast by special arrangement
Hotels (range/person/night)	Approximately $18 per day/person for half board
Other	Nil

PUBLICATIONS GIVING INFORMATION OF VALUE TO VISITORS

Laboratory can send, upon request, general information
on Tunisia as well as guides, lists of hotels, etc.

ADDITIONAL INFORMATION SUPPLIED BY THE STATION

The Mediterranean Marine Sorting Center offers a
service to the marine scientists of all countries
working in the Mediterranean area. New collections,
or collections that have been partly worked, are sent
to MMSC. Collections are sorted to specimen-groups
which are then distributed to specialists for study.
Identified series of specimens ultimately return to
MMSC, from where representative sets are returned to
the country of origin and to recognized depositories
around the world.

Further information supplied by writing:

                Director
                Mediterranean Marine Sorting Center
                Khereddine
                TUNISIA

NOTES

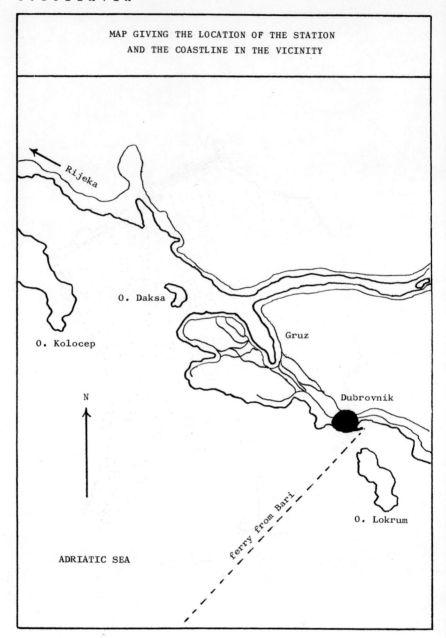

MAP GIVING THE LOCATION OF THE STATION
AND THE COASTLINE IN THE VICINITY

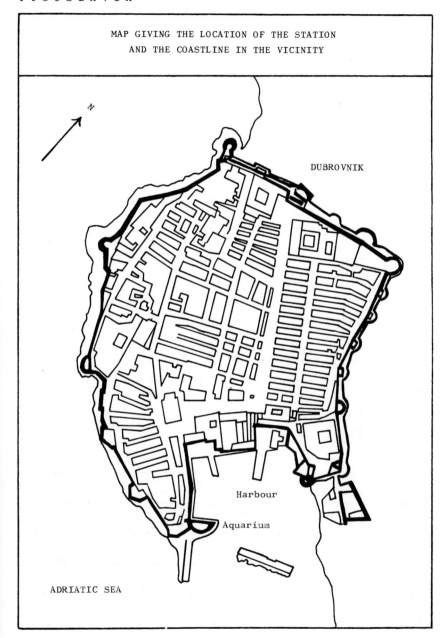

MAP GIVING THE LOCATION OF THE STATION
AND THE COASTLINE IN THE VICINITY

N

DUBROVNIK

Harbour

Aquarium

ADRIATIC SEA

Name and address of the Station	Biological Institute Yugoslav Academy of Sciences and Arts PO Box 39 Dubrovnik Yugoslavia
	Telephone 27-9.37
Director	Prof. Dr Tomo Gamulin
Affiliation	Yugoslav Academy of Sciences - Zagreb and University of Zagreb

NUMBER OF PERMANENT STAFF	
Scientific	5 marine biologists 1 ornithologist 2 botanists (land flora)
Technical	13 technical staff
Other	3 boat crew

ROUTES OF ACCESS	
Air	From Zagreb, Beograd, Rome (45 min)
Rail	From Zagreb and Beograd (24 hr)
Road	Adriatic coast road from Rijeka to Dubrovnik (400 Km)
Sea	Boat and ferry from Rijeka daily, from Bari 3 days a week

MAJOR ACCESSIBLE ECOSYSTEMS	
Intertidal	Rocky shore
Shallow sea	Narrow continental shelf
Ocean	Open sea beyond the shelf
Tidal range	About 50 cm

COLLECTING AND HYDROGRAPHIC FACILITIES	
on shore	Yes
from boats	Small boat 13 m 50 HP Various plankton nets for vertical and oblique hauls Nansen water-bottle, etc. Small trawl nets for bottom animals
Diving (including supply and/or refill of cylinders)	No possibility
Collecting service for live material provided by the Station	Common Mediterranean invertebrates provided for laboratory workers only. It is not possible to send specimens abroad
Hydrographic data	Records of temperature and salinity

SPECIAL ORGANISMS AND LOCALITIES THAT ARE A FEATURE OF THE STATION		
name	abundance	season
Mediterranean echinoderms	abundant	all the year round
Anemonia sulcata, Actinia	abundant	all the year round
Ascidians: Phallusia, Halocynthia, Microcosmus	abundant	spring to autumn
Sepia	moderate	March - May
Octopus	moderate	April - June
Morray eels	abundant	all the year round
Scylliorhinus canicula	moderate	spring to autumn

LABORATORIES OPEN TO VISITORS	
General laboratories	1 laboratory for zooplankton research and teaching (6 places)
Physiology laboratories	——
Biochemical laboratories	——
Research rooms for visitors	——
Wet sorting room	——
Experimental aquarium	Moderate facilities for different invertebrates
Controlled environment rooms	——
Photographic dark room	Yes
Workshop for general repairs to apparatus	Yes

LABORATORY SERVICES	
Electricity - voltage, DC/AC, frequency, phase	220 V AC 50 Hz single and 3 phase
Gas	Butane
Running seawater	Yes
Compressed air (pumps)	Yes
Computer link	——
General glassware	Yes
Chemicals	——
Balances (analytical)	1
Balances (torsion)	——
Refrigerators	Yes
Ovens	200 °C
Microscopes (compound)	——
Microscopes (dissecting)	2 (10-90 X)
Electron microscope	——
Centrifuges	——
Special apparatus	——
Library	Moderate, no journals.

LIVING ACCOMMODATION FOR VISITORS AND THEIR FAMILIES	
At the Station	None
Restaurant	None
Local Hotels	Many
Camp Sites	Yes
Car parking	Yes, but very restricted in the tourist season

APPROXIMATE CHARGES	
Laboratory	None
Boat	Usually not charged
Specimens	Usually not charged
Living accommodation at the Station	——
Meals at the Station	——
Hotels (range/person/night)	$6 - 20 per person full-board
Other	——

PUBLICATIONS GIVING INFORMATION OF VALUE TO VISITORS

The Laboratory will send a City of Dubrovnik Guide,
a list of lodgings and other information on request

ADDITIONAL INFORMATION SUPPLIED BY THE STATION

As the research of the station is solely in the field
of zooplankton only zooplanktologists can be accepted
to work as visitors in the laboratory

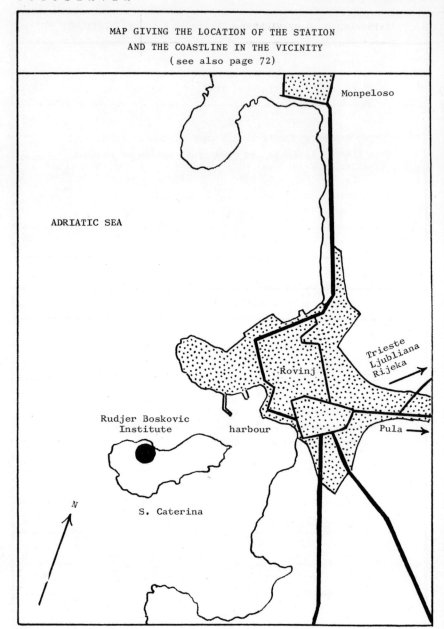

MAP GIVING THE LOCATION OF THE STATION
AND THE COASTLINE IN THE VICINITY
(see also page 72)

ADRIATIC SEA

Monpeloso

Rovinj

Trieste
Ljubliana
Rijeka

Rudjer Boskovic
Institute

harbour

Pula

N

S. Caterina

YUGOSLAVIA

Name and address of the Station	Center for Marine Research 'Rudjer Bošković' Institute 52210 Rovinj Yugoslavia		
	Telephone 052-81061 052-81029	Cables	Aquarium Rovinj
Director	Dr. Stjepan Kečkes		
Affiliation	Yugoslav Academy of Sciences & Arts, Zagreb and University of Zagreb		

NUMBER OF PERMANENT STAFF	
Scientific	At Rovinj: 18 biologists, 8 chemists, 2 physicists At the Center's laboratories in Zagreb: 22 scientists, not all involved in marine research
Technical	At Rovinj: 14 laboratory technicians
Other	At Rovinj: 6 crew, 3 administrators, 5 aquarium staff, 5 in the workshop, driver, cleaning personnel

ROUTES OF ACCESS	
Air	To Pula with a 50 min bus journey to Rovinj To Trieste with a 3 hr journey to Rovinj
Rail	From Ljubljana, Zagreb or Trieste to Kanfanar followed by 20 min bus journey to Rovinj
Road	32 Km from Pula, 105 Km from Trieste
Sea	Hydrofoil from Venice, May to October only

MAJOR ACCESSIBLE ECOSYSTEMS	
Intertidal	Exposed and sheltered rocky shores within 100 Km Limited mud or gravel locally
Shallow sea	Open sea with extensive continental shelf
Ocean	___
Tidal range	30 - 80 cm

COLLECTING AND HYDROGRAPHIC FACILITIES	
on shore	Simple shore collecting gear (nets, baskets) 1 combi-bus (driver available)
from boats	26 m RV 'Vila Velebita' with crew of five Plankton nets, bottom beam trawl, dredges (Riedl, Charcot, triangular, 'mušular'), grabs (Petersen and Van Veen), Dohrn water samplers, Nansen reversing bottles, 3 laboratories for hydrographic, primary productivity and other research at sea, piped hot and sea water, 220 and 380 V AC 50 Hz 24 and 12 V DC
Diving (including supply and/or refill of cylinders)	2 compressors 150 Atm (refilling possible) 8 cylinders (not available on loan) Personal equipment (not available on loan) 15 qualified divers Note: visitors require special permit for diving activity
Collecting service for live material provided by the Station	Common invertebrates and fishes supplied at the cost of sampling No live material sent abroad
Hydrographic data	Continuous records of temperature and salinity since 1924. Recent data on nutrients, microconstitutents, radioactivity. Unpublished data available to collaborating guest-scientists only

SPECIAL ORGANISMS AND LOCALITIES THAT ARE A FEATURE OF THE STATION		
name	abundance	season
Bryozoa (Hippodiplosia, Myriozoum, Cellaria etc.)	abundant	all the year round
Tunicata (Phallusia, Micro-cosmus, Ascidia, Distoma, Amaroucium)	abundant	all the year round
Bivalvia (Lima, Arca, Mytilus, Ostrea, Lithophaga)	abundant	all the year round
Aplysia rosea, A. depilans	moderate	spring - summer
Loligo vulgaris	moderate	summer
Branchiostoma lanceolatum	rather rare	summer
Actinia, Anemonia, Parazoanthus	abundant	warm seasons

LABORATORIES OPEN TO VISITORS	
General laboratories	2 laboratories (25 & 35 places) for student courses
Physiology laboratories	Open only to collaborating guest-scientists
Biochemical laboratories	Open only to collaborating guest-scientists
Research rooms for visitors	——
Wet sorting room	——
Experimental aquarium	Small and large tanks available
Controlled environment rooms	Under construction, open only to collaborating guest-scientists
Photographic dark room	Two; one of them open only to collaborating guest-scientists
Workshop for general repairs to apparatus	Yes, with 1 mechanic, 1 electrician, 1 carpenter

LABORATORY SERVICES	
Electricity - voltage, DC/AC, frequency, phase	220 V AC 50 Hz, single and three phase (stabilized 220 V AC current)
Gas	Butane
Running seawater	In some rooms
Compressed air (pumps)	Piped supply only in aquarium, individual pumps
Computer link	Not at present; desk calculators available
General glassware	Various
Chemicals	Common and analytical grades
Balances (analytical)	Single and double pans, simple and semi-automatic
Balances (torsion)	Several in the range 0.01 - 100 g
Refrigerators	Several refrigerators and deep-freezers
Ovens	Several, up to 1000°C
Microscopes (compound)	A few simple available on loan
Microscopes (dissecting)	A few (10-100X) available on loan
Electron microscope	——
Centrifuges	Several up to 15,000 rpm
Special apparatus	A wide variety of analytical and electronic equipment: polarographs, autoanalyzer, spectrophotometers, fluorimeter, single and multichannel analyzers, anticoincident low level beta counter, autoclaves, water baths
Library	Main marine journals, some available from 1890. Reports of expeditions. 12,000 vols.

Y U G O S L A V I A

LIVING ACCOMMODATION FOR VISITORS AND THEIR FAMILIES	
At the Station	No living accommodation or restaurant at the station
Restaurant	A wide choice locally
Local Hotels	9 hotels A, B and C categories (some open also out of tourist season) It is possible to rent houses, flats or rooms
Camp Sites	Two within 3 - 5 Km
Car parking	Near the station

APPROXIMATE CHARGES	
Laboratory	10 DM daily for laboratories for student courses. 2.50 DM daily per student
Boat	60 DM per hour
Specimens	No charge for common animals
Living accommodation at the Station	—
Meals at the Station	—
Hotels (range/person/night)	15 - 25 DM full board
Other	Private accommodation 5 DM per day per person

YUGOSLAVIA

PUBLICATIONS GIVING INFORMATION OF VALUE TO VISITORS

Lists of lodgings and prices will be sent on request
from the Center or from tourist companies 'Jadran',
'Turist', 'Kompas', 'Globtour'.

The Center's journal 'Thalassia jugoslavica' is
published regularly and covers all the fields of
marine science.

An extensive list of the marine flora and fauna has
been published by A. Vatova (1928) in: R. Com. Talass.
It., Mem. 143, 614pp.

ADDITIONAL INFORMATION SUPPLIED BY THE STATION

The Center for Marine Research at Rovinj is a
department of the 'Rudjer Bošković' Institute, Zagreb.

Although,in principle, no visitors can be accepted,
scientists from abroad are encouraged to enter into
active collaboration with the staff of the Center.
For collaborating guest-scientists all the facilities
of the Center are at their disposal and they are
charged only for the limited expenses the Center has
to meet in connection with their work.

Laboratories for student courses are run as a non-profit
making activity of the Center. About 500 students
(most of them from abroad) spend 10-15 days annually
at the Center.

Formal applications from prospective collaborating
guest-scientists should be addressed to the Deputy
Director of the 'Rudjer Bošković' Institute in Rovinj.
Applications for the use of laboratories for student
courses should be addressed to the Center's Secretariat.

NOTES